NCRP REPORT No. 138

Management of Terrorist Events Involving Radioactive Material

Recommendations of the NATIONAL COUNCIL ON RADIATION PROTECTION AND MEASUREMENTS

Issued October 24, 2001

National Council on Radiation Protection and Measurement
7910 Woodmont Avenue / Bethesda, Maryland 20814-3095

LEGAL NOTICE

This Report was prepared by the National Council on Radiation Protection and Measurements (NCRP). The Council strives to provide accurate, complete and useful information in its documents. However, neither the NCRP, the members of NCRP, other persons contributing to or assisting in the preparation of this Report, nor any person acting on the behalf of any of these parties: (a) makes any warranty or representation, express or implied, with respect to the accuracy, completeness or usefulness of the information contained in this Report, or that the use of any information, method or process disclosed in this Report may not infringe on privately owned rights; or (b) assumes any liability with respect to the use of, or for damages resulting from the use of any information, method or process disclosed in this Report, *under the Civil Rights Act of 1964, Section 701 et seq. as amended 42 U.S.C. Section 2000e et seq. (Title VII) or any other statutory or common law theory governing liability.*

Library of Congress Cataloging-in-Publication Data

National Council on Radiation Protection and Measurements.
Management of terrorist events involving radioactive material : recommendations of the National Council on Radiation Protection and Measurements.
p. cm. -- (NCRP report ; no. 138)
ISBN 0-929600-71-1
1. Radioactive substances -- Safety measures 2. Terrorism -- Prevention
RA569.M358 2001
616.9'897--dc21 2001052182

[For detailed information on the availability of NCRP publications see page 242.]

Preface

The National Council on Radiation Protection and Measurements has a long history of interest and concern over the radiation protection aspects of the accidental or intentional release of radioactive materials. This Report deals with releases of radioactive materials as the result of a deliberate act of terrorism. As so vividly demonstrated in the heinous attack on September 11, 2001, preparation for responding to such events must take place without regard to the specific nature of the event. This Report contains information and recommendations on radiation effects, medical management of radiation victims, issues surrounding the psychosocial impact, communications with the public and the media, and detailed recommendations on the organization and training of those responsible for responding to a terrorist event involving radioactive material. This Report was prepared prior to September 11, 2001.

The Council wishes to acknowledge the contributions made by many individuals who generously provided materials and/or briefed the Committee while this Report was being prepared. Besides providing essential background information and unique perspectives to the wide range of problems related to the radiation protection issues addressed in the Report, these individuals made many invaluable suggestions that were carefully considered by the Committee. James Fairobent (U.S. Department of Energy) provided current federal documents of relevance to the Committee's task and answered questions the Committee members posed related to the statement of work. Tom Dahlstrom (Bechtel, Nevada) provided a current overview of radiological threat scenarios. Anna Bachicha (U.S. Department of Energy) briefed the Committee on current DOE response capabilities including the current organizational structure and missions of DOE relative to nuclear weapons incident response. Susan Voveris (U.S. Department of Defense), the Commandant of the Defense Nuclear Weapons School, provided the Committee with an example of the type of modern training resources available for emergency response personnel as well as an example of interagency cooperation. Marion "Spike" Bowman (Federal Bureau of Investigation) provided an overview of the FBI's current strategy to manage domestic terrorism. Hank Austin (Texas

Task Force 1) provided invaluable information from the perspective of someone with responsibility for responding to a major incident at the state level. Rick Lane (Federal Emergency Management Agency) briefed the Committee on the role of the Federal Emergency Management Agency and the impact that the bombing of the Oklahoma City Federal Building had on the Federal Response Plan. Ron Williams (National Domestic Preparedness Office) described the organizational structure of the National Domestic Preparedness Office, their role in assisting local and state agencies prepare to deal with terrorist incidents, and recent initiatives to coordinate the response capabilities of the federal agencies. Beverly Ramsey (Los Alamos National Laboratory) provided information regarding current DOE initiatives concerning radiological incidents that may involve lands under the control of Native Americans. Tom Seed (U.S. Department of Defense) provided information regarding current and future medical countermeasures against exposure to life-threatening levels of ionizing radiation. The cities of Atlanta, Miami and New York generously provided the Committee with copies of their plans for dealing with radiological incidents.

This Report was prepared by Scientific Committee 46-14 on Radiation Protection Issues Related to Terrorist Activities that Result in the Dispersal of Radioactive Material. Serving on the Scientific Committee 46-14 were:

Chairman

John W. Poston, Sr.
Texas A&M University
College Station, Texas

Members

Cheri Abdelnour
Defense Threat Reduction Agency
Dulles, Virginia

E. John Ainsworth
Bethesda, Maryland

Steven M. Becker
University of Alabama at Birmingham
Schools of Public Health and Medicine
Birmingham, Alabama

Robert L. Brittigan
Defense Threat Reduction Agency
Dulles, Virginia

Ian Scott Hamilton
Texas A&M University
College Station, Texas

Eva E. Hickey
Battelle, Pacific Northwest Laboratories
Richland, Washington

David A. Kelm
Illinois Department of Nuclear Safety
Springfield, Illinois

Fred A. Mettler, Jr.
University of New Mexico School of Medicine
Albuquerque, New Mexico

Jay M. Thompson
Westinghouse Safety Management Solutions, Inc.
Aiken, South Carolina

Mark C. Wrobel
Bolling Air Force Base
Washington, D.C.

NCRP Secretariat

Eric E. Kearsley, *Consultant*
Cindy L. O'Brien, *Managing Editor*

The Council wishes to express its appreciation to the Committee members for the time and effort devoted to the preparation of this Report and to the U.S. Department of Energy (DOE) for the financial support provided to enable the NCRP to complete this effort.

Charles B. Meinhold
President

Contents

1. Introduction

Throughout the history of civilization, sizable segments of the population have suffered from the occurrence of many different types of major disasters including earthquakes, hurricanes, tornadoes, famines and plagues. Today, in the United States, and in many other countries, local, state and national governmental organizations are maintained specifically to respond to such disasters. In modern society, these disasters also elicit an outpouring of responses by numerous nongovernmental organizations having charters for much wider services but who also operate at the local, national, and even international levels. Usually, the individuals comprising these organizations possess a spectrum of talents and training designed to prepare them to respond to a variety of natural and human-made emergencies.

Over the last few decades, there have been several events worldwide that involved the potential or actual dispersal of radioactive materials. These events make it clear that issues involving radiation and radioactive materials must be addressed in proper emergency response plans prepared to address terrorist threats.

In the past, the National Council on Radiation Protection and Measurements (NCRP) has offered advice for specific types of radiological emergencies. For example, NCRP Report No. 42 (NCRP, 1974) addressed radiological factors influencing decision making in a nuclear attack. This report provided some guidance for use in a large-scale disaster involving an intense and uncontrolled exposure of many people to ionizing radiation. This report superseded NCRP Report No. 29, *Exposure to Radiation in an Emergency* (NCRP, 1962). In direct response to the accident at Three Mile Island (TMI), the NCRP issued Commentary No. 1 addressing the public health significance of releases of ^{85}Kr into the atmosphere (NCRP, 1980a). Later, the NCRP hosted a symposium on the control of public exposure in the event of a nuclear incident[1] or an

[1]Although the word "incident" is often used to describe events of relatively small scale, in this Report "incident" is used without regard to the magnitude of the event.

attack (NCRP, 1982). The theme of this symposium was the protection of the general population against the radiation consequences of war, terrorism or accident. Finally, the NCRP published Commentary No. 10, *Advising the Public about Radiation Emergencies: A Document for Public Comment* (NCRP, 1994). This commentary reviewed the salient features in providing the public with information regarding radiation emergencies.

Recent events involving terrorist activities both within the United States and abroad have focused attention on the level of our preparedness to deal with large-scale radiological, chemical and biological threats. The full spectrum of radiological threats from terrorists spans the deliberate dispersal of radioactive material to the detonation of a nuclear weapon. While the most likely threat is the dispersal of radioactive materials, the use of a crude nuclear weapon against a major city cannot be dismissed. Even though the effects of ionizing radiation have been well studied and documented, especially when compared with most chemical and biological threats, there remains a need for additional radiation safety guidance for emergency planners and emergency responders, including those responsible for restoring the disaster area.

1.1 Purpose of this Report

This Report provides information and recommendations regarding the radiological health and safety issues related to the threat of terrorist activities involving radioactive material. The Report identifies, evaluates, and makes recommendations regarding immediate and long-term radiological consequence-management issues, communication and coordination challenges, and public information challenges associated with these emergencies. This Report also provides recommendations on training guidelines, critical resources, and guidelines for internal and external exposure, as well as decontamination and cleanup. The NCRP is aware of several existing recommendations and plans from many levels of government within the United States. Rather than reiterate this information (which is constantly evolving), our attention has been focused on the basic principles underlying effective planning and response to terrorist activities associated with the dispersal of radioactive materials. The Scientific Committee that prepared this Report reviewed the most likely radiological-threat scenarios taking into consideration release and dispersal mechanisms, lessons learned from actual disasters and large-scale radiological exercises, and current disaster plans at the local, state and federal levels.

1.2 Target Audiences of this Report

While most of this Report is applicable regardless of where in the world the event occurs, certain portions (*e.g.*, discussions of local, state and federal response plans) are focused on an event occurring in the United States. The Report should be useful to individuals at all levels of government, including tribal governing bodies, as well as nongovernmental bodies who share the responsibility to respond to this type of disaster. While the Report is aimed at this wide audience, the focus is on those things that would be most useful for first-responders, hospitals, and specialists in radiation safety who are likely to be included in the response.

The Report is presented in a number of sections addressing a spectrum of radiation safety issues related to terrorist activities. Each section is structured to stand alone and addresses very specific issues; as such it should be useful to the specialist in a particular area. The entire Report would be useful as a reference document for individuals responsible for emergency planning as well as in the training of individuals who will be involved in emergency response operations.

1.3 Scope of this Report

The Report is organized into three broad areas: the definition of the problem (Sections 2 to 5), management of the disaster (Sections 6 to 9), and preparation for a major radiological incident (Sections 10 to 12).

Section 2 provides a frame of reference for the range of issues that are involved in the management of this type of incident as well as the challenging factors that complicate the decision-making process. A discussion of the devices and the physical consequences of their use is presented in Section 3. Health effects associated with exposure to ionizing radiation, medical management of exposed and/or contaminated individuals, and human services associated with these emergencies are discussed in Section 4. The immediate and long-term psychosocial impact of terrorist activities, an important but often neglected subject, is discussed in some detail in Section 5.

Section 6 provides a discussion of the command and control issues in terms of the responsibilities and authorities vested in the local, state and federal authorities likely to respond to such an emergency. Included in the series of appendices to this Report is information on current federal, state, and local plans for dealing with major radiological disasters. Section 7 addresses the critical issue of communications with the public and the media. It includes recommendations for managing the flow of information including logistics issues, information policy, clearance authority, ethical issues, and the importance of risk communication with the public. Additional material related to public communications is provided in other appendices contained in this Report. The first is a checklist of suggested early public communications actions and the second is a list of questions and answers that may arise from the public/media concerning terminology and risks associated with radiological disasters. Section 8 presents an overview of appropriate guidance for dose limitation and cleanup after an incident. Recommended exposure guidelines and cleanup criteria are presented in this Section. Radiological consequence-management considerations are presented in Section 9 of the Report. These considerations include management during the early, intermediate and late phases of the emergency and include discussions of recommended protective and recovery actions.

Section 10 addresses emergency planning and the critical resources, technical assets, and equipment needed to respond to these types of emergencies. Training and qualifications are discussed extensively in Section 11, with additional supporting information presented in an appendix. Finally, the Report concludes with a summary of recommendations for future research (Section 12) and a summary of the major recommendations related to responses to these types of emergencies (Section 13).

1.4 Acronyms

The use of acronyms in a report of this nature is unavoidable. Although each acronym is defined when it is first introduced, a list is provided following the Glossary.

2. Considerations Impacting Response

The purpose of this Section is to provide a frame of reference in which to consider the issues involved in responding to terrorist activities that result in the actual or potential dispersal of radioactive materials.

2.1 Unique Features and Potential Impacts

A terrorist act involving dispersal of radioactive materials is qualitatively different from conventional terrorist acts involving explosives.[2] In a conventional terrorist act, the event or series of events occurs, casualties are suffered, and survivors are rescued and treated. The trauma experienced by the victims is familiar—cuts, broken bones, burns, etc. Those not killed or injured in the immediate terrorist act are free from further physical danger. Psychological shock and horror are likely to result from the attack, but at least the attack is bounded in time and space. While the debris from such acts may be extensive and present well-understood hazards to those responding (*e.g.*, fires, structural instability, sharp edges, etc.) the materials are not inherently hazardous and the cleanup is localized. The immediate site of the event can be secured as a crime scene and forensic investigation can be conducted in the usual manner. It is unlikely that a conventional terrorist event could prevent state and local authorities from providing normal government services.

When an explosive device is used to disperse radioactive materials, the paradigm shifts. Treatment of casualties is more difficult because of the contamination and the complications associated

[2]Many of the characteristics of terrorism involving radioactive materials are also present in terrorism involving other weapons of mass destruction such as chemical and biological materials. However, the focus of this Report is terrorist activities that involve the potential or actual dispersal of radioactive materials.

with other trauma. There is a real potential for physical injury and death to persons who were not wounded in the immediate terrorist attack. The debris from the event and other normally harmless materials will be contaminated. The affected area may be much larger than the immediate scene of the crime. The radiological threat, invisible and uncertain in terms of long-term health impacts, will engender considerable public fear and concern. The incident will be difficult to manage until appropriate monitoring equipment and well-trained technical individuals are available. Forensic investigation will be complicated by the need to wear protective equipment, the contamination of evidence, and the pressure to cleanup or stabilize the crime scene. Finally, there is a broad range of potential effects on critical infrastructure.

Conceptually, there are three classes of events. The least serious is a localized spill that can readily be controlled and decontaminated by local or state personnel with the appropriate training and equipment. This type of occurrence does not present significant public policy concerns and generally does not require outside radiological assistance. At the other extreme is an event causing such extensive damage that local and state authorities, emergency, fire and medical services are overwhelmed and cannot provide normal government services. Detonation of a device with even a small nuclear yield in a major metropolitan area could have this effect. In this case, the federal government would have to provide such government services and might have to temporarily assume control until such time as local and state government could resume normal functions. In between, there is a broad range of possible occurrences in which local and state governments continue to function but will require assistance from federal authorities. A range of potential scenarios are discussed in Section 3.

2.2 Factors Complicating the Decision-Making Process

In responding to any emergency, it is important to understand what decisions need to be made, what information is required to support those decisions, and what the possible consequences, both good and bad, of any decision may be. In the context of a radiological terrorist event, there are also many potentially confounding factors.

2.2.1 *Law Enforcement Interests*

Unlike response to an accidental fire or explosion resulting in the release of hazardous materials, public health and safety is not the sole focus of responders to a terrorist event. Instead, law enforcement and other security related concerns must also be considered. For example, with regard to decisions related to information release, authorities must consider the possible effects on (1) gaining and maintaining control of the incident (*e.g.*, before a device is detonated, knowledge of actions being contemplated by authorities may prompt terrorists to carry out an attack); (2) identifying and capturing other possible participants and accomplices; and (3) successfully prosecuting the cases in court. In deploying resources, decision makers must consider the possibility that responders may be attacked, that other devices may have been introduced, that there may be chemical or biological materials as well as sources of ionizing radiation, and that other attacks may take place at other sites. Decisions on how best to resolve the situation will have to consider such legal and constitutional issues as preservation of evidence, maintenance of chains of custody, and limitations on the legal authority to conduct searches. These legitimate law enforcement interests are not necessarily incompatible with prevention or mitigation of the threat to the public; however, they may complicate the decision-making process.

2.2.2 *Public Health and Safety*

Depending on the nature of the radionuclides involved and the manner and extent of dispersal, the radioactive materials may present an immediate threat, a long-term threat, or both. Decision makers will have to be concerned with protection of the response forces, decontamination and treatment of casualties, and protection of the general public. Immediate availability of expertise, specialized equipment, and supplies may all be limiting factors. The ability to enforce recommendations for sheltering or evacuation will depend on public participation as well as local and state laws. In many states, there is no statutory authority to enforce a recommendation to evacuate. Restrictions on food and water will have both public health and economic implications. Some short-term precautions may give rise to unanticipated and unnecessary long-term consequences such as loss of tourism or markets for products. Long-term considerations include public health, psychosocial effects, environmental remediation, and economic impacts. With all of these potentially competing interests, it will be important to assure that decisions are both timely and well informed.

2.2.3 *Mass Casualties and Damage to Infrastructure*

Use of an improvised nuclear device, stolen weapon, or large radiological dispersal weapon in an urban environment will cause chaos. Depending on the nature and size of the device, the location in which it is used, and environmental conditions, the number of casualties will vary dramatically, with medical management potentially compromised by either the large numbers of casualties or by the complexity of injuries that are the result of the combined effects of radiation exposure, trauma, burns, or other insults. The ability to deal with mass casualties and minimize further loss of life will be complicated by the collateral effects of the device's blast, ensuing fires, and widespread contamination. Electromagnetic effects from a nuclear weapon may disrupt communications systems, electrical distribution networks, computers, and other technology. Blast and fire can destroy or block normal transportation routes, hindering access to casualties. Similarly, severe fallout or deposited radioactive materials may preclude rescue attempts. Under such conditions, the command and control authority and the entire emergency response team must continue to function, making the best effort to provide support. Although it is likely that in the early stages these efforts will not be entirely successful, it is clear that with perseverance and the support of the nation and the international community, order can be restored and services provided to all who require support.

2.2.4 *Psychosocial Impacts*

In preparing for, or responding to, terrorist incidents involving radioactive releases, it is crucial to recognize the centrality of social and psychological issues. At the most basic level is the fact that one of terrorism's chief aims is psychological: to induce fear in a population. Such fear is further compounded when "invisible toxins" such as radiation are involved. People can neither see nor sense the presence of radiation, but they know that it is potentially hazardous. Because the threat cannot be perceived with the unaided senses, and because of frightening historical associations (*e.g.*,

Hiroshima, Nagasaki, Chernobyl), radiation incidents have a powerful potential to create fear and dread. Under such circumstances, a critical challenge facing the responsible authorities is to develop a communication strategy that is informed by an awareness of people's fears and concerns and that effectively conveys the information needed to protect health and safety.

At the same time, it is important to bear in mind that the significance of the psychosocial component is far broader than the matter of fear. Indeed, some of the most difficult aspects of consequence management after a radiological release may relate to the social and psychological aftereffects of an incident. A substantial body of research conducted over the past two decades makes it clear that the experience of contamination (or even possible contamination) can have profound psychosocial impacts on individuals, families and communities. While some of these impacts are acute in nature, others can be more chronic. At the individual level, for example, this can mean elevated levels of distress many years after the initial incident, while at the community level, evacuations or relocations necessitated by contamination can have serious and long-lasting social impacts. Efforts to prevent, mitigate and ameliorate such impacts will be an important component of consequence management.

Also requiring attention will be the problem of social stigma. In the aftermath of radiological incidents, the affected areas, and even the persons who come from these areas, can come to be seen by others as "tainted." In other cases, use of terminology (*e.g.*, "radiophobia") can be perceived as being dismissive of real health concerns. This can significantly complicate efforts to address the consequences of an incident. Finally, the task of reestablishing trust and a sense of safety after an incident will be a daunting but essential part of managing the long-term consequences of a terrorist-related radiological incident. In sum, psychosocial issues—in the short, medium and long term—will be an important part of any scenario involving the release (or possible release) of radioactive materials. As such, consideration of social and psychological factors will need to be an integral part of preparedness and response efforts.

2.2.5 *Environmental Concerns*

Except for scenarios involving very short-lived radioisotopes, the dispersal of radioactive material will necessarily result in a requirement for decontamination and remediation as well as possible long-term monitoring. Although the radionuclides may differ, the environmental problems associated with radiological terrorism are similar to those that might result from a nuclear incident. Actions taken in the early and mid phases of the response are likely to have a profound impact on site restoration activities. In addition to health concerns, responsible officials will have to take into consideration the restoration of confidence of local residents, potential economic partners, and customers upon whom their future economy will depend. Disposal of the radioactive waste will be controversial. In the event of a major dispersal, the costs of recovery will become an important factor. While the long-term nature of the cleanup effort will permit time for considered decision making, the issues will be very complex.

2.3 Functional Approach

Effective response to a terrorist event resulting in the potential or actual dispersal of radioactive materials will require a wide variety of skills and expertise. In addition to the law enforcement and possible military assets necessary to respond to the terrorists, such an incident will require radiological, medical and other expertise to address the radiation issues. Because of the nature and possible magnitude of terrorist events involving radioactive materials, there may exist unique requirements for protective actions, mass care, disaster relief, public health, mental health, and public affairs expertise. The requirement to integrate and effectively use these varied resources will give rise to complex command, control and management challenges. These are addressed in Section 6.

In recognition of the different skills involved, it is useful to consider the response to a terrorist event as being composed of two principal functions: crisis management and consequence management. While the boundaries between the two are not always clear, the concept is helpful in planning and executing the response.

2.3.1 *Crisis Management*

The crisis-management portion of the response is focused on the terrorists and on preventing their intended actions, *i.e.*, counterterrorism. The Federal Bureau of Investigation (FBI, 1998) has described crisis management as follows:

Crisis management addresses the **causes** of a terrorist incident—the identity, motivation, and capability of the terrorists and the weapons they employ. Crisis management is a law enforcement function and includes measures to identify, acquire, and plan the use of resources needed to anticipate, prevent, and/or resolve a threat or act of terrorism. In a terrorist incident, a crisis-management response may include traditional law enforcement missions, such as intelligence, surveillance, tactical operations, behavioral assessment, negotiations, forensics, and investigations, as well as technical support missions, such as agent identification, search, render safe procedures, transfer and disposal, and limited decontamination. In addition to the traditional law enforcement missions, crisis management also includes assurance of public health and safety.

2.3.2 *Consequence Management*

As its name implies, consequence management is focused on the results of the incident and may be required over a long-term period. The FBI (1998) has defined consequence management as follows:

Consequence management addresses how the incident **affects or potentially might affect** public health, safety, and the environment. Consequence management includes measures to protect public health, safety, and the environment, to restore essential government services, and to provide emergency relief to governments, businesses, and individuals affected by the consequences of terrorism.

The relationship between crisis and consequence management is summarized in Figure 2.1.

The theoretical defining moment between crisis management and consequence management is the successful, or partially successful, execution of the terrorist act. An incident may consist primarily of crisis-management operations. When there is no warning prior to a terrorist attack, *e.g.*, the April 19, 1995 Oklahoma City bombing, consequence management may predominate during the early phases of the response (*e.g.*, rescue of the injured), and crisis management may predominate during later phases (*e.g.*, recovery of forensic evidence). In most cases, both crisis management and consequence management will occur simultaneously before, during and after the dispersal of radioactive material.

Command and control relationships are affected by the distinction between crisis management and consequence management. This is addressed in more detail in Section 6.

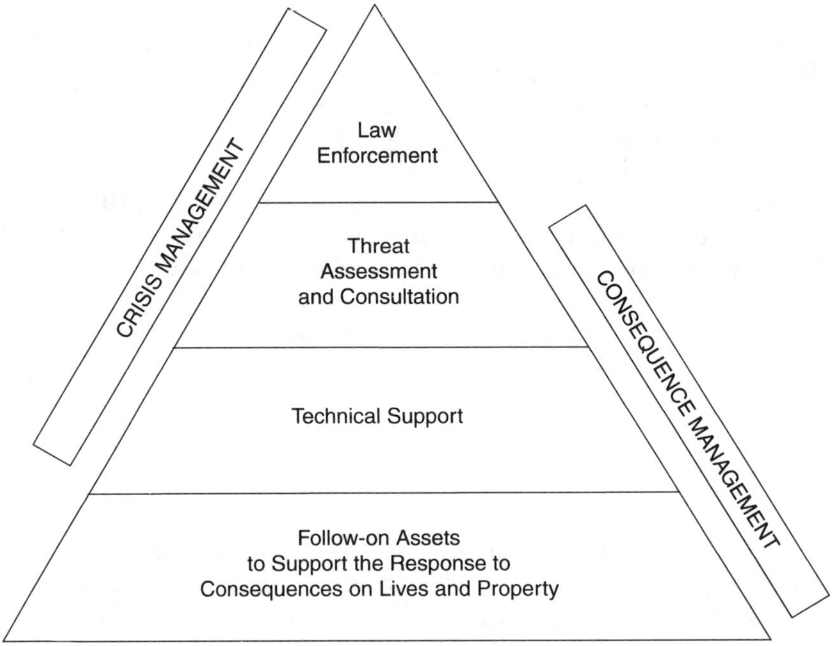

Fig. 2.1. The relationship between crisis management and consequence management [adapted from Figure TI-1, Terrorist Annex to the Federal Response Plan (FEMA, 1999)].

3. Characteristics and Consequences of Terrorist Incidents that Involve Radioactive Materials

The purpose of this Section is to broadly describe the characteristics and likely consequences of two general categories of terrorist incidents that could have widespread radiological consequences. The first is the use of conventional explosives or other mechanisms to disperse radioactive materials and the second is the use of nuclear weapons. In addition to a radiological dispersal device (RDD), the first category also includes the intentional dispersal of radioactive material resulting from an attack on fixed nuclear facilities or radioactive material in transit as well as malfunctioning nuclear weapons that are detonated with no nuclear yield (a nuclear "dud").

The second category of terrorist incidents considered in this Report is the use of nuclear weapons. These weapons might be constructed from nuclear material and conventional explosives or they might be stolen from military stockpiles. The detonation of a weapon with even a small nuclear yield will cause significant radiological consequences in addition to substantial damage to the infrastructure. These consequences result from both the initial ionizing radiation at the time of detonation and from radioactive fallout that will occur for a considerable time after the initial event.

3.1 Radiological Dispersal Events

For convenience and clarity, radiological dispersal incidents are divided into two broad categories: those involving small and generally highly localized sources and those involving the dispersal of large amounts of radioactive materials over large areas.

3.1.1 *Localized Sources*

A single or a few, small low-level (containing small amounts of radioactive material) sources may be used with the principal objective of causing fear within a population and ultimately of disrupting the social order. The radioactive material could be packaged in a small container such as an ampule, shoe box, or even a suitcase-sized container. If in liquid form, the material could be dumped into a water reservoir or spilled over some small area; or, to create mayhem over a larger area, it could be released in small amounts from a bicycle, motor vehicle, or even an aircraft. Because the amount of radioactivity is small, the exposure to individuals would also be expected to be low. Thus, the harm from this kind of source is primarily psychosocial, and whatever low external or internal dose is received should produce no immediate adverse health effects and only a small probability of long-term health effects. Health effects consequences from exposures to low levels of ionizing radiation are discussed in Section 4 and psychosocial effects in Section 5.

While the principal route of exposure is external, some internal contamination could occur if the radioactivity is inhaled or enters the food chain. For a well-localized event this would be treated like the spill of any hazardous material. Protective clothing will prevent or at least help to minimize the contamination of emergency responders. However, it will not be possible to shield against penetrating radiation and care should be taken to minimize the time spent in close proximity to high concentrations of the material. Generally these sources are easily located with the use of radiation detection instrumentation and effective protective measures designed to control the source and limit exposures to the public may be taken very quickly. Detailed guidance is provided for medical personnel in Section 4 and for other emergency response personnel in Section 9.

3.1.2 *Widely Dispersed Sources*

Of greater concern are events that result in the dispersal of radioactive materials over large areas through the use of explosives coupled with large amounts of radioactive material. If the target area is populated, individuals injured by the explosion are likely to be contaminated with radioactivity. Greater amounts of radioactive materials would likely be used in such devices and radiation casualties may include individuals who could have received life-threatening levels of exposure. The objective of such a device is similar to that of a smaller source, but is intended to affect an extended area or population.

The most likely scenarios involve the use of a solid radioactive material that would be of low enough activity that the construction and delivery of the RDD will not seriously inhibit the terrorist from carrying out the attack. Large sources of penetrating radiation are difficult to handle safely and without detection by authorities. Shielding materials that are adequate to protect both the individuals who construct these devices and those who are to deploy them complicate the design and fabrication of effective weapons. Although not insurmountable, these challenges can only be overcome with considerable technical expertise and sophisticated resources.

Although the most likely devices involve a high explosive coupled with a solid, usually pellets or powder, the radioactive material could also be in some kind of solution, or even be a radioactive gas.

The area over which these materials will be dispersed depends on the amount of explosive, atmospheric conditions, and the extent to which the radioactive material adheres to dust or other material dispersed by the explosion. Any gases will escape, but finely dispersed radioactive particulates, or just chunks of metal, will contaminate the ground and the surfaces of structures. In this scenario, it is most likely that only a small area of a few city blocks would be involved, but like a chemical spill, care is needed to avoid the spread of the material into other areas. As before, it is expected that most exposures would be low and the principal health and psychosocial effects in the aftermath of such an event would be similar to those discussed in Section 3.1.1, but for a larger population.

Nuclear reactors, adjacent spent fuel storage depots, nuclear fuel reprocessing facilities, transport vehicles, or any high-level waste site are potential targets for the use of high explosives to disperse into the atmosphere the very high levels of radioactivity associated with materials at these facilities. A successful incursion into a nuclear power reactor would require a very heavily armed force, since commercial reactors are very well protected. Only when a reactor is being refueled and the containment structure is open would atmospheric dispersion of the reactor's nuclear fuel be likely as a result of the use of high explosives. However, after reactor shutdown, less radioactive material is contained in the fuel than during normal operations because short-lived fission products quickly decay to low or negligible levels. Also, because there is less decay heat in the fuel, there is less energy to drive fission products out of partially damaged fuel.

Spent nuclear fuel elements could also be targeted, but they contain much less radioactive material than an operating reactor plant because of the rapid decay of fission products.

Concerning the affected area, health hazards expected from dispersal of highly radioactive nuclear fuel would be similar to that which occurred at Chernobyl, but on a significantly smaller scale. Radioactive gases, liquids and particulates would serve as sources of both external and internal exposure. Within the containment structure, exposure rates could be high enough to result in lethal doses within a matter of hours. With increasing distance, both exposures and other hazards would be lower. The areas at risk from high-level radioactive waste dispersed by a large explosive device can be many miles from the source. With a smaller amount of source material and explosive, the area of concern is more in the range of several city blocks or a few miles from the target area.

The inventory of radioactive material at research reactors in universities or other facilities is very small in comparison to that in power reactors. Therefore, the anticipated impact from such an attack would also be significantly less.

3.2 Nuclear Weapons

In contrast with the civil defense scenarios of the Cold War involving an exchange of large numbers of high-yield weapons, the most likely terrorist nuclear weapons scenario involves the use of a single, probably low-yield device. Although catastrophic, the availability of resources from the state, the federal government, and even the international community make the consequences of this type of disaster manageable.

In this Section, the basic characteristics of low-yield nuclear weapons will be summarized with an emphasis on the information of greatest use to emergency planners. This information presents the ranges over which significant effects are likely to have an impact on people and structures. There are numerous other references such as Glasstone and Dolan (1977) and Northrup (1996) that treat this subject in much greater detail.

3.2.1 *Yield*

Judging from the yields of the earliest nuclear weapons and the technical difficulties of building nuclear weapons, it is probable that a subnational organization would be limited to the construction of a crude nuclear weapon of less than about 10 kilotons[3] (KT). However, acquisition of a compact, higher-yield weapon may be possible by stealing a stockpiled weapon. In this Section, effects will be described for nuclear weapons from 0.01 to 10 KT; if a higher-yield device is detonated, the types of effects will largely be consistent with, and consequences far in excess of the lower-yield cases.

The successful placement of a nuclear weapon in a city requires that the device be transported and placed in a covert manner. Placement of a nuclear weapon in a city would require that the device remain undetectable until the explosion; as the size, weight, and radiation signature increase, the probability of detection increases. A homemade weapon would likely be physically larger than military weapons that are constructed using advanced technology and manufacturing techniques. A stolen weapon would be much more compact, but safeguards in the device may preclude a nuclear yield from attempted detonation by an unauthorized party. All of these factors lessen the likelihood that a large nuclear device could be successfully placed and detonated in an urban area.

The lowest yield, 0.01 KT, is analyzed to provide perspective on a device that "fizzles" with some nuclear yield, as compared to a "dud" with no nuclear yield. It would seem unlikely that a device with a lower yield would be intentionally designed, since an equivalent yield using conventional explosives could be more easily constructed. It is noted however that a weapon with a yield of 0.01 KT, though described above as a "fizzle" still would have an impact much greater than the explosive that destroyed the Oklahoma City Federal Building on April 19, 1995.

3.2.2 *Effects*

The following summary of effects is meant only as semi-quantitative descriptions to illustrate the most significant effects of

[3]The special unit "kiloton" used in connection with nuclear yield refers to an equivalent amount of TNT. The conversion to SI units given by Glasstone and Dolan (1977) is that one ton of TNT is equivalent to 4.2×10^9 J.

nuclear weapons. The factors that influence the probability and magnitude of these effects are discussed. The yield/range relationships are taken from Glasstone and Dolan (1977), as implemented in the HOTSPOT (fallout prediction code) suite of computer codes (Homann and Wilson, 1995). While more sophisticated models are available, these estimates are useful in understanding the relationships between the most significant weapons effects.

3.2.2.1 *Air Blast*. As with a conventional explosive, a nuclear detonation produces a shock wave in air that propagates outward from the point of detonation. This shock wave, also referred to as a blast wave or an overpressure wave, is a transient pressure wave usually measured in pounds per square inch (psi).[4] This air blast and the accompanying strong winds can produce damage directly to structures and injuries to people. Injuries may also result indirectly from falling debris and missiles (*e.g.*, flying glass shards) produced by the interaction of the air blast with buildings and other structures. Window glass is especially vulnerable to damage and may break at pressures of less than 1 psi. Once broken, glass shards may be accelerated by the shock wave. With sufficient velocity, these glass shards can cause injuries and fatalities. In the following analyses, a 50 percent fatality rate from flying glass is assumed for persons close to windows at an overpressure of 12 psi. Radii derived for this fatality rate (Table 3.1) are very approximate, since orientation of the glass and the proximity of persons to windows will vary. Predicting the radii at which specific damage or injuries will occur is also complicated by the complex geometries present in an urban setting (multiple reflecting surfaces for shock wave propagation), and the weapon-yield dependent variations in the shape and duration of the shock wave profile. Injuries and other effects should be expected at greater radii than those listed for fatalities.

3.2.2.2 *Thermal Radiation*. Detonation of a nuclear device produces an extremely hot fireball, with temperatures peaking at tens of millions of degrees kelvin. The radiant energy from the fireball is sufficient to ignite materials and cause burns far from the fireball. The intense light associated with the fireball may also cause temporary or permanent blindness. Shadowing by structures between the fireball and potential receptors will prevent or reduce

[4]1 psi is equivalent to 6,985 pascal, the SI unit for pressure.

TABLE 3.1—*Approximate ranges for 50 percent fatalities from flying glass accelerated as a result of a shock wave.*

Yield (KT)	Range for 12 psi Overpressure (m)
0.01	60
0.1	130
1	275
10	590

thermal effects. Although the type and degree of injury caused by thermal radiation depends on a number of factors (duration of the thermal pulse, skin pigmentation, area of exposed skin, etc.) a thermal exposure[5] of 8 cal cm^{-2} is commonly assumed to be the value at which a 50 percent mortality from burns occurs. Table 3.2 provides a summary of ranges at which this thermal exposure occurs for various weapon yields.

3.2.2.3 *Initial Nuclear Radiation.* The detonation of a nuclear weapon produces an initial intense pulse of ionizing radiation. Both gamma rays and neutrons are released. The radiation produced in the first minute post-detonation is termed *initial radiation,* and that resulting from the decay of radioactive materials after the first minute is termed *residual radiation.*

TABLE 3.2—*Approximate ranges for fatalities from thermal radiation.*

Yield (KT)	Range for 50% Mortality from Thermal Burns (8 cal cm^{-2})(m)
0.01	60
0.1	200
1	610
10	1,800

[5]The calorie (cal) is a commonly used unit of energy. 1 cal is equivalent to 4.19 J.

The value of 4 Gy (absorbed dose in tissue measured at the body surface) is the approximate value for 50 percent mortality (LD_{50}) for fatality from acute exposure without medical treatment. This value corresponds roughly to a midline body absorbed dose of 3 Gy (Levin *et al*., 1992).[6] Lower doses may also be fatal for persons with other injuries (such as from the air blast or thermal burns). Approximate ranges at which an individual would receive an LD_{50} are provided in Table 3.3. Intervening buildings can provide some radiation shielding. Basements of buildings provide additional protection. Ranges corresponding to an acute exposure to 4 Gy should be considered approximate, and may vary depending on nuclear device design. For a given yield, absorbed dose increases rapidly as "ground zero" is approached for a given yield; the range at which 20 Gy is received is about 75 percent of the range for 4 Gy. Note that the absorbed dose versus range relationship does not obey a simple inverse-square relationship because of the interactions of the air with the complex, mixed radiation field from a nuclear weapon.

3.2.2.4 *Residual Nuclear Radiation*. Residual nuclear radiation is defined as the ionizing radiation that is emitted after the initial intense pulse of radiation from the detonation of a nuclear weapon. This includes the significantly increased levels near the site of detonation caused mostly by the radioactive weapon debris as well as the activation of soil and other materials by components of the initial radiation. A ground-level nuclear explosion causes the injection of vast quantities of these radioactive materials into the atmosphere, subsequently increasing the amount of fallout at all

TABLE 3.3—*Approximate ranges for a 4 Gy absorbed dose from initial radiation.*

Yield (KT)	Range for 4 Gy Due to Initial Nuclear Radiation (m)
0.01	250
0.1	460
1	790
10	1,200

[6]Note that with competent medical support, an individual could survive an acute exposure to ionizing radiation up to three to four times this midline absorbed dose (see Section 4).

distances. Fallout patterns are a function of the yield, height of burst, and meteorological conditions. The fallout calculation results presented in Table 3.4 assume a ground level detonation and typical meteorological conditions. Estimated residual absorbed doses for various weapon yields at three different radii are listed in Table 3.5. It is clear from this Table that approaching ground zero soon after detonation is extremely hazardous. Note that fallout will not be present for the entire hour since the radioactive material must first be transported downwind. For instance, in the example above, the absorbed dose at 9,600 m downwind for the 10 KT device is received in the last 24 min of the first hour, since it takes the cloud about 36 min to reach that location. This delay may give authorities time to issue protective action recommendations to the public before the fallout reaches downwind areas.

TABLE 3.4—*Ranges for an absorbed dose of 4 Gy from residual radiation in the first hour after detonation.*

Yield (KT)	Range for 4 Gy from Fallout in First Hour after Detonation (m)
0.01	1,270
0.1	2,750
1	5,500
10	9,600

TABLE 3.5—*Residual absorbed dose in first hour after detonation (Gy).*

Yield (KT)	1,000 m	2,000 m	10,000 m
0.01	6.7	1.5	0.02
0.1	38	8.3	0.1
1	210	47	0.6
10	1,200	260	3.5

3.2.2.5 *Crater Formation.* A nuclear weapon detonated near the surface of the ground will displace soil and form a crater. The crater size depends on the height above ground, the yield, and mechanical characteristics of the ground. Table 3.6 presents the approximate apparent radius of the crater formed by weapons of various yields for a ground level detonation in soft rock. These radii were derived from data and relationships presented by Glasstone and Nolan (1977). Crater radii vary greatly with the mechanical characteristics of the ground. In wet soil, a 1 KT ground burst will produce a crater with a radius of about 25 m. In dry, hard rock the radius is about 15 m. When detonated below, but near the surface, the radius varies rapidly with depth of burial.

3.2.2.6 *Ground Shock.* In addition to the air blast, a nuclear weapon detonated close to the ground will produce a ground shock. This ground shock can disrupt utilities and damage structures. The ground shock and air blast would be expected to cause major disruptions in the local infrastructure. Damage would not be repairable for some time after the detonation, perhaps for weeks or months. Isolation of the most heavily damaged areas may be necessary to restore proximate areas.

3.2.3 *Discussion*

The area affected for most of the described effects is essentially a large circular region centered on the detonation site. Thus, the range associated with a given effect can also be considered the radius of the circle where a given effect may be observed, neglecting any shadowing or intervening shielding. The only major effect that does not exhibit uniform azimuthal coverage about the detonation site is fallout. Fallout occurs primarily in the downwind

TABLE 3.6—*Radius of crater for a ground level detonation in dry soft rock.*

Yield (KT)	Radius (m)
0.01	5
0.1	10
1	20
10	40

direction from the nuclear detonation. However, because wind and precipitation patterns can be complex, the pattern of deposition of radioactive materials from fallout will also be complex.

Table 3.7 provides a summary of the ranges for several significant effects of a nuclear weapon detonation. Neglecting the more complex fallout pattern, the immediate effects of most importance are the initial nuclear radiation and thermal burns. For small nuclear explosions, a person in the line of sight receiving significant but nonlethal thermal burns may still receive a fatal dose of nuclear radiation. As yield increases to around 1 KT, thermal burns will begin to become as important as initial radiation for producing fatalities. At 10 KT, thermal burns will produce fatalities at a greater distance than the initial radiation. These relationships are illustrated in Figure 3.1.

An important conclusion from this discussion is that the spectrum of the types of casualties (burn, radiation, etc.) expected from a low-yield nuclear weapon will depend on the yield of the weapon, with radiation casualties being most important at the lowest yields. At the higher end of the range of yields being considered in this Report, casualties with multiple types of injuries are more likely to present themselves.

The contaminated cloud presents a hazard to personnel in aircraft (as well as to those on the ground). News and police aircraft would likely provide coverage of the event. Personnel in aircraft

TABLE 3.7—*Summary of ranges for significant effects.*

Yield (KT)	Range for 50% Mortality from Air Blast (m)	Range for 50% Mortality from Thermal Burns (m)	Range for 4 Gy Initial Nuclear Radiation (m)	Range for 4 Gy Fallout in First Hour after Blast[a] (m)
0.01	60	60	250	1,270
0.1	130	200	460	2,750
1	275	610	790	5,500
10	590	1,800	1,200	9,600

[a]Unlike the other effects listed in this Table, the pattern of deposition of radioactive materials from fallout depends on local meteorological conditions especially wind patterns and precipitation.

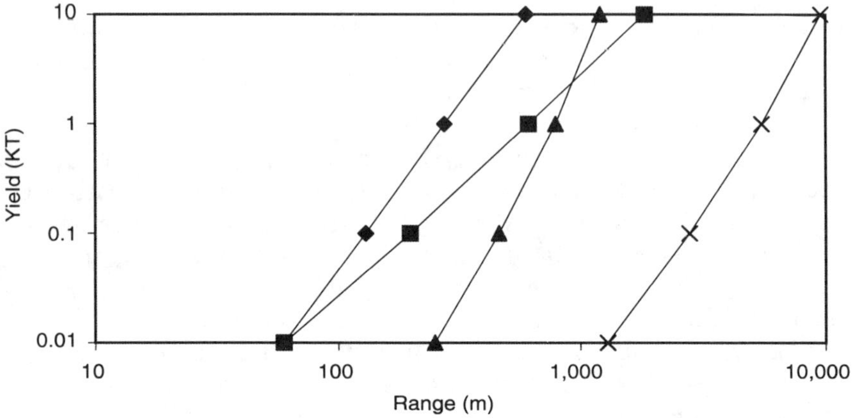

Fig. 3.1. Ranges for a 50 percent probability of fatality from nuclear weapons by different mechanisms. Although multiple injuries caused by the interaction of the various types of injury will increase the probability of fatality, this effect is not taken into account in this diagram (◆ = blast, ■ = thermal radiation, ▲ = initial radiation, and × = fallout).

that pass through or close to the cloud may receive significant absorbed doses.

The use of population densities, averaged over several square kilometers of land area may significantly underestimate the potential casualties from a low-yield weapon that may be targeted at a cluster of people (such as an office building during business hours, or a major sporting event). Even the smallest nuclear yield considered here, 0.01 KT, could cause tens of thousands of deaths from initial radiation effects if detonated at such a location. The radius of potentially lethal exposures resulting from the detonation of a very small (0.01 KT) nuclear device extends to over 200 m.

Table 3.8 provides a summary of the characteristics of the types of threats discussed in this Section. The Table outlines the magnitude of each threat and provides information about how each of these threats may be recognized by emergency response personnel.

TABLE 3.8—*Summary of radiation threats, magnitude of the event, and how each may be identified.*

Source/Kind of Material	Size of Source	Hazards	Evidence of the Event
Small radioactive source: *e.g.*, nuclear medicine and brachy-therapy sources, industrial gauges, small calibration sources	< several mm^3 to >100 cm^3	External radiation	Visible radiation symbol Lead shielding materials
Large isotopic source: *e.g.*, ^{192}Ir- for radiography, ^{60}Co for teletherapy	<1 cm^3 to many m^3, depending on configuration and presence of shielding materials	External radiation	Visible radiation symbol Lead shielding materials
RDD with large isotopic source	<500 cm^3 to >1 m^3, depending on configuration, shielding and dispersal mechanism	Blast, if explosives used, internal and external radiation	Unusual debris (*e.g.*, small metallic fragments and lead shielding) Potential luminescence Unexplained heat
RDD with spent nuclear fuel	>1 m^3, depending on configuration, shielding and dispersal mechanism	Blast, internal and external radiation	Unusual debris (rods and pellets, shielding) Potential luminescence Unexplained heat
Attack on commercial power reactor or reactor accident	Fuel rod assembly: ~5 m in length	External and internal radiation	Self evident
Attack on research reactor or research reactor accident	Similar to power reactors but much smaller quantity of fuel/ fission by- product material	External and internal radiation	Self evident

TABLE 3.8—*Summary of radiation threats, magnitude of the event, and how each may be identified.*

Source/Kind of Material	Size of Source	Hazards	Evidence of the Event
Transportation accident involving radioactive material	Cardboard packages, 30–55 gallon drums, low specific activity containers, shield pigs.	External, and if packaging breached, internal radiation	Visible radiation symbols Radiation placards
"Dud" nuclear weapon	Backpacks to much larger containers	Blast, internal radiation	Unusual debris High levels of alpha radiation Unexplained heat
Nuclear weapon	Backpacks to much larger containers	Blast, thermal, external and internal radiation	Blinding flash, severe heat, and blast waves Grit/sandy fallout

4. Medical Management of Radiation Casualties

The goal of this Section is to describe the adverse health effects associated with exposure to ionizing radiation as a result of the types of terrorist events discussed in Section 3 and to describe the medical management of patients with radiation injuries. Guidance is provided for on-scene responders and for medical personnel in hospitals who will treat these patients.

4.1 Basic Terminology

Human exposures to ionizing radiation can be either *external*, when the source of radiation is outside the body, or *internal*, when radioactive material enters the body. External exposures may be large as in the case of an accidental exposure from a food irradiator or a planned therapeutic exposure from a radiotherapy source such as a medical accelerator. It may also be a small exposure as in the case of a dental or medical x ray, or the cosmic-ray component of natural background. External exposure may be penetrating (*e.g.*, neutrons, x or gamma rays) or nonpenetrating (*e.g.*, alpha or beta particles).

Radioactive material can enter the body by eating or drinking contaminated food or fluids, through skin or a wound, and by breathing radioactive gases or aerosols. Emergency responders should wear protective gear (see Section 9) and this should virtually eliminate the inhalation or ingestion of radionuclides, and careful decontamination will greatly decrease the radiation exposure of contaminated individuals. A radioactive material taken into the body will distribute through physiological processes determined by its chemical and physical properties. Radioactive materials that remain on the surface of the skin are considered sources of external exposure if they do not enter the body. However, while on the skin, they can be inhaled, ingested or enter the body through a break in the skin.

A variety of terms are used to describe and categorize the effects of ionizing radiation. First, the time period over which effects are manifested can be described by the terms *early* and *late*. Early effects generally refer to the consequences of the exposure that are expressed within a period of a few days to a few months. Late effects refer to the long-term consequences of the exposure and include effects that may not be expressed for many years. A second set of terms, *acute* and *chronic* are used to describe the period of time over which an individual is exposed. Generally an acute exposure refers to an exposure received within a period of a few hours or less and a chronic exposure generally refers to exposures received over several days or longer. It is important to recognize that both acute and chronic exposures can give rise to both early and late effects.

The biological effects of ionizing radiation can be categorized as being either *deterministic* or *stochastic*. Deterministic effects are those that are assumed to have a threshold (*i.e.*, an exposure level below which the effect is not observed) and whose severity increases with the exposure level. In contrast, stochastic effects are those that are assumed not to have a threshold and whose severity does not depend on the exposure level. Skin reddening (erythema) is an example of a deterministic effect because it has a threshold of approximately 5 to 6 Gy with increasing severity for larger exposures.[7] Skin reddening is also an early effect because it is usually expressed within two to four weeks after the exposure. Cancer caused by ionizing radiation is an example of a stochastic effect because it is assumed to have no threshold and because the severity of the cancer, once it occurs, is independent of the exposure. Cancer is also an example of a late effect because there is a long period of time, usually many years, between the exposure and the expression of the disease.

The quantities and their associated units used in the radiological sciences are divided into two categories: those considered fundamental and those that are derived from the fundamental quantities for specific applications (*e.g.*, radiation protection). The most important fundamental quantity used in this Report is *absorbed dose*, defined as the quotient of $d\bar{\varepsilon}$ by dm where $d\bar{\varepsilon}$ is the mean energy imparted by ionizing radiation to the matter in a volume element and dm is the mass of the matter in that volume

[7]An earlier skin reddening response can occur at lower absorbed doses (2 Gy) within a few hours of the exposure. This is caused by damage to the superficial capillaries and generally resolves within 2 d.

element, *i.e.*, the absorbed dose $D = \mathrm{d}\bar{\varepsilon}/\mathrm{d}m$. The unit of absorbed dose is the joule per kilogram (J kg^{-1}) and is given the special name, gray (Gy).

A quantity derived from the absorbed dose and used for radiation protection purposes is the *dose equivalent*, defined as the product of a dimensionless quality factor, and the *absorbed dose* to tissue at a specific point. A quality factor is chosen to weight the absorbed dose by the biological effectiveness of the charged particle spectrum at the point in tissue where the absorbed dose is determined (ICRU, 1993). The unit of *dose equivalent* is the joule per kilogram (J kg^{-1}) and is given the special name, sievert (Sv).

To emphasize the importance of the *absorbed dose* averaged over a tissue or organ in radiation protection, the International Commission on Radiological Protection (ICRP) introduced the quantity *equivalent dose* defined as the product of the *absorbed dose* averaged over the tissue or organ of interest and a radiation weighting factor appropriate for the type of radiation incident on the body that resulted in the *absorbed dose* to the tissue. The unit of *equivalent dose* is the joule per kilogram (J kg^{-1}) and is given the special name, sievert (Sv).

The special units described above (gray and sievert) are a part of the International System [Le Système International d'Unités (SI)] of Units. The NCRP (1985) has endorsed this system and recommends its use in the United States. However, the conventional unit for *absorbed dose* (rad), and the conventional unit for the *dose equivalent* (rem), continue to be used widely. A summary of the conversions between these units is provided at the end of this Report in a section following the Glossary.

4.2 Spectrum of Health Effects

Beginning with a brief description of radiation effects at the cellular level and progressing through the most important early and late effects of ionizing radiation, this Section provides a brief overview of the most important biological effects of radiation.

4.2.1 *General Considerations: Cellular Damage and Absorbed Dose Rate*

The health effects of ionizing radiation depend largely on the absorbed dose, the absorbed dose rate, and the organs or tissues that have been exposed. Radiation damage to the cell's genetic material [deoxyribonucleic acid (DNA) and mitotic apparatus] can cause cell death or, if damaged cells survive, can result in altered cell or tissue function. For example, death of bone marrow stem cells can result in low platelet, white and red blood cell counts and, consequently, a high susceptibility to infection and bleeding. Damaged DNA in surviving cells can cause mutations in the cells and an increased risk of cancer.

The absorbed dose rate has an important impact on radiation damage and personnel hazard. Cells have the capacity to repair injury to their genetic material and, at low absorbed dose rates, these repair mechanisms can decrease the frequency of lethal and nonlethal injuries to the cell. This is an extremely important consideration for emergency response personnel because of the impact on risk and because most exposure scenarios confronted by emergency response personnel following the initial event involve protracted exposure. More practically, the absorbed dose rate will determine the amount of time a person may remain in an area without incurring unacceptable long-term risks of adverse health effects. It is essential that before emergency response personnel enter an area of elevated exposure, the anticipated exposure be justified in terms of the objectives to be accomplished. In addition, these personnel should be provided with radiation detection equipment that can be used to assess personnel exposures, identify types of radiation, and establish the boundaries of the contaminated areas.

4.2.2 *Early Effects of Ionizing Radiation in Humans*

Nonlife-threatening effects include temporary or permanent sterility, depression of rapidly proliferating cell types (*e.g.*, bone marrow stem cells), vomiting, skin reddening, hair loss, and cataracts. Table 4.1 provides estimates of acute exposure thresholds for these effects. In general, thresholds are higher if the exposure is protracted over periods of time greater than a few hours.

The acute radiation syndrome is a broad term used to describe a range of signs and symptoms that reflect severe damage to specific organ systems and that can lead to death within hours or up to several months after exposure. The nature of these injuries, the time at which they are expressed, and often the duration are a function of the absorbed dose and the rate at which it is received by the individual. The separate component syndromes such as the hematopoietic syndrome, the gastrointestinal syndrome, the

TABLE 4.1—*Estimated threshold absorbed doses*[a] *for deterministic radiation effects following an acute exposure (adapted from IOM/NRC, 1999a).*

Exposure Health Effect	Organ	Absorbed Dose (Gy)
Temporary sterility	Testis	0.15
Nausea		0.35
Depression of blood cell forming process	Bone marrow	0.5
Reversible skin effects (*e.g.*, early reddening)	Skin	2
Permanent sterility	Ovaries	2.5 – 6
Vomiting		3
Temporary hair loss	Skin	3 – 5
Permanent sterility	Testis	3.5
Skin erythema	Skin	5 – 6

[a]The absorbed doses reported in this Table refer only to exposures to low-LET radiation (*i.e.*, x rays, gamma rays, or energetic electron beams).

cardiovascular syndrome, and the central nervous system (CNS) syndrome are discussed in detail in a number of references (Conklin and Walker, 1987; Mettler and Upton, 1995). The following description of the symptoms associated with a large, acute exposure to ionizing radiation is taken from the *Medical Management of Radiological Casualties Handbook* prepared by the Armed Forces Radiobiology Research Institute (AFRRI, 1999).

Whole-body gamma absorbed doses as low as 0.35 Gy can cause nausea, weakness and appetite loss within a few hours following acute exposure. These symptoms will disappear within a few hours. In the range of 0.70 to 1.25 Gy, there is a 5 to 30 percent probability of transient nausea within a few hours of the exposure. Death is not expected unless the individual is more susceptible to the effects of the exposure because of other factors such as chronic infection. Between 1.25 and 3 Gy, there is an increasing probability and intensity of nausea and vomiting as well as mild to moderate weakness. If exhibited, these symptoms will persist for up to 2 d.

Although mortality is still expected to be low in this exposure range, medical problems will include infection, bleeding and fever. Wounds or burns received by patients will significantly increase morbidity and mortality. Midline absorbed doses to the whole body in the range of 3 Gy (4 Gy tissue absorbed dose free-in-air) produce about a 50 percent chance of death for the untreated adult individual within about 60 d (Levin *et al.*, 1992). Studies in experimental animals indicate juveniles may be more susceptible. At higher absorbed doses, signs and symptoms of lower absorbed doses will persist and become more severe. In addition, within 3 or 4 d, the patient will experience frequent diarrhea, anorexia, increased fluid loss, ulceration, and the probability of death increases dramatically. At absorbed doses in excess of 5 Gy without treatment, mortality could reach 100 percent with some deaths occurring within as little as two to three weeks (AFRRI, 1999). With effective treatment (see Section 4.4.3), individuals suffering acute radiation injuries with no other complicating factors are likely to recover at absorbed doses up to 10 Gy.

In cases of external exposure, nonuniform or partial-body irradiations can occur when part of the body is shielded from the source or when the exposed individual is close to the source. For example, contamination of the unprotected skin with radionuclides by close contact with radioactive material from an RDD or from fallout may produce extensive localized damage to skin and underlying tissues. For radioactive material taken into the body, the absorbed dose distribution will depend on the route of intake and the kinetics of the radionuclide determined by its chemical and physical properties. Some radionuclides are distributed essentially uniformly in body tissues (*e.g.*, ^{137}Cs), whereas other radionuclides are preferentially deposited in certain organs or tissues (*e.g.*, ^{131}I in the thyroid).

4.2.3 *Radiation Carcinogenesis*

Cancer, including leukemia, has been clearly linked with exposure to ionizing radiation, and is likely the most important effect at absorbed dose levels below 1 Gy. Over the last five decades, thousands of papers on radiation carcinogenesis have appeared in the scientific literature. The most careful analysis and review of this field is provided in periodic reports of the United Nations Scientific Committee on the Effects of Atomic Radiation (UNSCEAR, 2000) and the U.S. National Research Council Committee on the Biological Effects of Ionizing Radiation (NAS/NRC, 1990).

Ionizing radiation can induce either benign or malignant tumors which are generally described as stochastic effects. These are effects without an assumed threshold and for which increasing the absorbed dose to the individual increases the probability of a cancer, but has little or no effect on its severity.

Contrary to public perception, ionizing radiation is a relatively weak carcinogen. As an example, among the approximately 86,000 atomic-bomb survivors at Hiroshima and Nagasaki who have been studied from 1950 to 1990, there has been an excess of only 334 deaths from solid cancer (7,578 versus 7,244 expected) and there have only been 87 excess deaths from leukemia (249 versus 162 expected) (Pierce *et al.*, 1996). Epidemiology studies of these populations continue and a small excess of cancers and some other diseases is still being detected more than 50 y after exposure, especially among people who were young when irradiated. Although many types of human tumors can be induced by radiation exposure, the sensitivity of specific tissues to cancer induction by radiation varies significantly, and a few types of neoplasms do not appear to be radiation induced (*e.g.*, chronic lymphocytic leukemia). Once a radiogenic tumor occurs in a given individual, it is clinically and pathologically indistinguishable from tumors due to other causes.

Radiation-induced cancers are characterized by a *latent period*, that is, the elapsed time between radiation exposure and the clinical appearance of the disease. Minimum latent periods are 2 to 3 y for leukemia, 3 to 4 y for bone cancer, 4 to 5 y for thyroid cancer, and approximately 10 y for the other solid tumors. If tumors are found in a very short time interval after radiation exposure, causes other than ionizing radiation must be seriously suspected. Risk of radiation-induced solid cancer has been shown to persist for 40 y or more, and may persist for a lifetime, although limited follow-up makes this uncertain for those exposed early in life.

Data on the carcinogenic risk from moderate exposures are available from epidemiological studies and generally show an increasing risk with increasing absorbed dose. Several studies involving large therapeutic absorbed doses (several gray) to specific organs have suggested a decline in risk at very high doses. The decline in incidence of tumors at very high doses is thought to reflect the death of cells that are damaged and could have caused cancer had they survived.

Several mathematical dose-response models have been proposed for radiation-induced carcinogenesis. The simplest of these is often referred to as the *linear-nonthreshold risk model*. With this model, there is an increase in the probability of cancer that is directly proportional to the absorbed dose.

At small, incremental absorbed doses above background, the linear-nonthreshold risk model usually cannot be excluded or definitively confirmed; the small number of radiogenic cancers that are expected at low doses may be concealed by the large number of spontaneous cancers. Because of this statistical limitation, directly inferring risk at low doses may never be possible using population studies. Recent mortality analyses of the atomic-bomb survivors suggest that for solid tumors, a linear dose response is consistent with the data for acute absorbed doses as low as 0.05 Gy (Pierce and Preston, 2000; Pierce *et al.*, 1996). The dose-response relationship for leukemia observed in the 1996 study appears to be curvilinear, with an increasing slope as the absorbed dose increases until it plateaus above 1.5 to 2 Gy, presumably due to cell killing.

For radiation protection purposes, the NCRP (1993a; 1993b) has endorsed nominal values of the lifetime risk of fatal cancer of 10×10^{-2} Sv^{-1} for a population of all ages exposed at a high dose and high-dose rate. At low doses or for protracted exposures, the Council assumes that the risk of a fatal cancer varies linearly with absorbed dose, without threshold, and that this risk coefficient is to be reduced by a factor of two to take into account the normal recovery capabilities of the body. The NCRP has performed an analysis of the uncertainties associated with this risk estimate and concluded that the 90 percent subjective confidence interval (5th to 95th percentile) ranges from 1.2×10^{-2} Sv^{-1} to 8.84×10^{-2} Sv^{-1} (NCRP, 1997). The radiation protection quantity with which these nominal risk coefficients should be applied is the effective dose. However, risks to specific individuals should be based on all available information including the radiation type and quantity, absorbed doses to specific organs, as well as the appropriate age- and tumor-specific risk factor.

4.2.4 *Effects of In Utero Irradiation*

It is possible that pregnant females may be exposed to radiation or to radionuclides during a terrorist incident. As a result, some information on potential effects on the fetus are included here. Fetal dose from external radiation can be generally estimated through a knowledge of the type of radiation, its penetrating power and some estimate of the beam direction and skin or maternal surface dose. If the absorbed dose to the uterus can be calculated, it provides a reasonably close estimate of embryonic or fetal absorbed dose. In a situation in which radionuclides are inhaled or ingested by the mother, the absorbed dose to the fetus depends on the physical and chemical nature of the compounds. Compounds that are very soluble in water and are in ionic form (such as iodides) can easily cross to the fetus through the placenta, whereas with other compounds, the placental barrier provides protection of the fetus. A recent NCRP report (NCRP, 1998) provides detailed information on most radionuclides of interest.

Development of the unborn child may be divided into three major phases. These include: (1) the preimplantation phase from conception to implantation, (2) the phase of major organogenesis which extends from the third to approximately the eighth week, and (3) the phase of fetal development lasting from nine weeks until birth which includes the major period of CNS development from the 8th to the 25th week.

Absorbed doses to the fetus in the range of 0.1 Gy to several gray can result in significant fetal harm. Above a practical threshold, damage from ionizing radiation during pregnancy that results in cell killing or unrepaired damage can cause a wide range of identifiable abnormalities, including lethality, CNS abnormalities, cataracts, growth retardation, malformations, and even behavioral disorders. Since the *in utero* neural system is most sensitive, neuropathology will always accompany other abnormalities in humans.

The effects of radiation on the fetus depend on the time of exposure relative to conception. When the number of cells in the embryo is small and their nature is not yet specialized, the effect of damage to these cells is most likely to take the form of failure to implant or of an undetectable death of the fetus. Exposure of the embryo in the first three weeks following conception is not likely to result in effects in the live-born child, despite the fact that the CNS and the heart are beginning to develop in the third week. During the rest of the period of major organogenesis, conventionally taken to be from the third week after conception, malformations may be caused in the organ under development at time of exposure. These effects appear to have a threshold of at least 0.1 Gy.

During the period of 8 to 25 weeks, the CNS is relatively sensitive to radiation. Fetal absorbed doses in excess of 0.1 Gy may result in a decrease of intelligence quotient (IQ). During this same period, fetal doses in the range of 1 Gy result in a high probability of severe mental retardation. Values of IQ lower than expected have been reported in some children exposed *in utero* at Hiroshima and Nagasaki. There have been two principal quantitative findings. The first is reduction in IQ with increasing absorbed dose. This effect is very dependent on fetal age. Regardless of the time of gestation, IQ reduction has not been clinically identified at fetal absorbed doses of less than 0.1 Gy. In the period from 8 to 15 weeks after conception a fetal absorbed dose of 1 Gy reduces IQ by about 30 points. A similar, but smaller, reduction is detectable following exposure in the period from 16 to 25 weeks.

At fetal absorbed doses of 1 Gy the probability of this effect is about 40 percent. The effects of all levels of dose are less marked following exposure in the period from 16 to 25 weeks after conception and have not been observed for other periods. All the clinical observations on IQ and severe mental retardation relate to high absorbed dose and high absorbed-dose rates and their direct use for chronic exposures probably overestimates the risks.

Throughout most of a pregnancy, the fetus is assumed to be at risk for potential carcinogenic effects of radiation. From the third week after conception until delivery there is felt to be an increased risk of both leukemia and childhood cancer. The magnitude of the risk has been the subject of many publications, yet their interpretation remains open to debate. Doll and Wakeford (1997) have shown elevated risks associated with obstetric x-ray examinations of pregnant women. Fetal absorbed doses associated with that study were about 10 mGy. There is some evidence of elevated numbers of leukemias among atomic-bomb survivors who were irradiated *in utero* but there is no apparent dependence on absorbed dose and the cases did not occur during childhood.

Fetal absorbed doses in the range below 0.1 Gy appear to present no substantial risk of fetal death, malformation or impairment of mental development. In addition, the lifetime risk of radiogenic induction of childhood cancer or leukemia at 0.1 Gy is about 1 in 170. Accordingly, the ICRP (2000) has concluded that there is no medical justification for terminating a pregnancy at fetal absorbed doses below this level.

4.2.5 *Other Late Effects*

Other late effects of concern include (1) severe genetic (hereditary) effects expressed in subsequent generations and (2) other causes of noncancer mortality associated with exposure to ionizing radiation. The irradiation of the gonads of either parent has not been shown to result in deleterious effects on children. Over the last three decades, it has become clear that the risks of transmitting such radiation-acquired abnormalities to offspring have been difficult or impossible to identify. No hereditary effects have been seen in human studies below gonadal absorbed doses of 0.5 Gy. At higher absorbed doses, there are very few populations of large enough size to allow risk estimation. As a result, human risk estimates have been based largely on analyses of animal data. The NCRP has endorsed a risk coefficient for severe genetic effects of 1×10^{-2} Sv^{-1} for a population of all ages exposed to low absorbed doses and absorbed-dose rates (NCRP, 1993b).

The category "other causes of noncancer mortality" includes diseases of the circulatory, digestive and respiratory systems. Statistically significant increases in mortality attributable to these diseases have been observed in the atomic-bomb survivors. Although there are insufficient data to determine a dose-response relationship, the current data appear to show a curvilinear shape with essentially zero risk below 0.5 Sv. The relative increase in the mortality rate for these diseases for individuals exposed to 1 Sv is approximately 10 percent (Shimizu *et al*, 1999).

4.3 Medical Management of Radiation Casualties

This Section addresses the on-scene management of radiation casualties, the treatment of patients who have received a significant whole-body exposure, and also patients who have inhaled radioactive material or who have wounds involving radioactive materials. The psychosocial problems associated with a radiological event are treated in Section 5.

4.3.1 *On-Scene Triage*

Treatment of life-threatening injuries always takes precedence over measures to address radioactive contamination or exposure. Individuals with such injuries should be stabilized, if possible, and immediately transported to a medical facility. If it does not delay the medical response, an individual with radiation protection training should accompany the patient to provide radiation protection assistance. This individual is an advisor to the medical team and is subordinate to the senior medical person responsible for the patient. The possibility of contamination on or in the patient may be determined in the field, en route to a treatment facility, or at a hospital depending on the condition of the patient. The hospital that will receive these patients should be informed of the number of patients, the nature of their injuries, and whether or not they are suspected of being contaminated.

Other injured personnel should be sorted and treated according to standard medical triage guidelines with the exception that those who are contaminated should be separated so that they can receive a preliminary decontamination (see Section 4.3.2) before or during transport to a hospital for final treatment.

The symptoms of individuals who have received large absorbed doses of radiation include nausea, vomiting, fatigability and weakness. The symptoms exhibited by individuals who have received large absorbed doses are described in greater detail in Section 4.2.2 and summarized in Appendix A. These symptoms can also reflect an exposure to many toxic materials and are also reported by some who experience great psychological stress. Because of the large absorbed doses required to cause these symptoms, it is unlikely that they are caused solely by radioactive contamination that may be present on the patient.

Patients who have no evidence of external contamination, but are likely to have internal contamination as a result of a wound, an inhalation or ingestion of radioactive materials, may be treated in routine medical or emergency rooms. However, blood, vomitus, urine or feces may be contaminated and should be handled with care.

Patients with large amounts of radioactive material imbedded in a wound warrant special attention because activated metal can contain radionuclides with very high specific activities and there may be a significant exposure hazard to treatment personnel. Dose equivalent rates from such fragments may be as high as 1 Sv h^{-1} very close to the object. Such incidents could occur from an explosion in the reactor of a nuclear power plant or from an RDD.

Individuals who are only externally contaminated and not otherwise injured should preferably be decontaminated at some place other than a hospital. Taking such persons to a hospital will divert needed medical resources from critically injured patients.

4.3.2 *Patient Radiological Assessment*

The radiological assessment of an injured individual should be performed by an individual with radiological health training and only under the supervision of on-scene medical personnel. This assessment includes radiation measurements and collection of information that is relevant to the decontamination and treatment of the patient. The instrument used to perform the survey should be sensitive to both penetrating and nonpenetrating radiation (*e.g.*, a Geiger-Mueller tube with a thin wall or entrance window). Care should be taken not to contaminate the probe by contact with the patient or any other potentially contaminated surface. If the patient is in a contaminated area, the individual should be moved to an area of lower background under the supervision of the senior medical person on the scene. The distribution of radioactivity should be recorded for each patient along with other relevant remarks such as the location of wounds. Administrative information such as the patient's name; the name of the individual conducting the survey; the time, date and location at which the survey was performed; and the serial number and type of instrument used should be recorded. A survey form with a diagram of an anatomical figure such as Standard Form 531 available from the U.S. General Services Administration web site (http://www.gsa.gov/forms/medical.htm) is suitable for this purpose.

Examples of the type of information that may be helpful in the early medical management of radioactively contaminated persons and that could be collected by medical/radiological health personnel at the scene or during transport to the hospital are listed below:

Circumstances of the incident:

- When did the terrorist event occur and what are the circumstances of the incident?
- What are the most likely pathways for exposure?
- How much radioactive material is potentially involved?
- What injuries have occurred?
- What potential medical problems may be present besides the radionuclide contamination?
- What measurements have been made at the site of the incident (*e.g.*, air monitors, smears, fixed radiation monitors, nasal smear counts, and skin contamination levels)?
- Are industrial, biological or chemical materials involved in addition to the radionuclides?
- Have any treatments been given for these?

Present status of the patient:

- If known, what radionuclides now contaminate the patient?
- Where and what are the radiation measurements at the surface?
- Was the patient also exposed to penetrating radiation? If dosimetry information is available, what has been learned

from processing personal dosimeters, *e.g.*, film badge, thermoluminescent dosimeter, or pocket ionization chamber? If not yet known, when is the information expected?
- What information is available about the chemical and physical properties of the compounds containing the radionuclides (*e.g.*, solubility, particle size)?
- What decontamination efforts, if any, have already been attempted? With what success?
- What therapeutic measures, such as the use of blocking agents or isotopic dilution procedures have been taken?

Patient follow up:

- Has clothing removed at the site of incident been saved?
- What excreta have been collected?
- Who has the samples?
- What analyses are planned?
- When will they be done?

In addition to being useful for the treatment of the patient, this information will be valuable to the On-Scene Commander (OSC). Good communication between medical personnel and other components of the initial response team is particularly critical in the early phases of the response.

4.3.3 *Personnel Decontamination Procedures*

This Section applies only to contaminated individuals with no other significant injuries and to contaminated patients under competent medical supervision.

Internal contamination is a much greater problem than external contamination because it is often difficult to remove and residence times within the body may be very long. Therefore, external decontamination procedures are designed to minimize or prevent internal contamination of the patient and those providing care. Radionuclides on the intact skin surface rarely cause a high enough absorbed dose to be a hazard to either the patient or to medical staff.

The external decontamination process begins with the single most effective action: the removal of the outer clothing of the contaminated individual. This usually will remove most surface contamination. The clothing should be placed in a sealed container (*e.g.*, a plastic bag). Each container should be labeled with the patient's name, location, time and date, and marked clearly with: "RADIOACTIVE — DO NOT DISCARD." These items may be analyzed later to identify the radionuclide or perhaps to perform a particle size analysis that would be useful to assess internal exposure from an inhalation. After removing the contaminated clothing, if inhalation is suspected, a nasal sample, from both nostrils, using two clean swabs can be taken for later analysis. Depending on the extent of contamination and the medical condition of the patient, a full body shower may then be used.

For a more localized area of contamination, a simple irrigation may be all that is needed. Tepid water, with or without a mild detergent is generally very effective. Hot water is not used in order to avoid a hyperemia that may increase absorption of contaminants through the skin. Cold water is also not used since it would tend to close skin pores and trap radioactive contamination. The decontamination of intact skin should begin with areas of highest contamination levels and progress to areas of lower contamination levels. Every effort should be made to avoid contamination of otherwise clean areas. Because the intact skin is a very effective barrier to internal contamination, every care should be made to avoid irritating or in any way compromising this barrier. Procedures such as shaving or harsh scrubbing are not appropriate.

Although it is usually not required, if hair needs to be removed, clipping is effective. Decontamination should begin with the least aggressive method and progress to more aggressive ones, always taking care not to break or irritate the skin. Radioactive material removed from the patient should be preserved for later analysis to identify the specific radionuclide.

Complete decontamination is generally not possible because some radioactive material can remain fixed to the skin surface. Decontamination efforts that lower contamination levels to a level twice background are usually considered adequate. In any case, decontamination efforts should be reevaluated or suspended as soon as the survey instrument reveals that no further progress is being made. Additionally, it should be noted that on-scene decontamination should be only as thorough as is practical under the circumstances under which it is being performed. For example, for individuals who will ultimately be transported to a medical facility, the final evaluation of the radiological status of the patient can be made at that location. Finally, following the completion of decontamination, a record of the radiological status of the individual should be made. This record should contain information such as that described in Section 4.3.2.

A summary of these on-scene procedures is provided in Table 4.2. Following completion of the preliminary decontamination, the patients who need further care are ready for transport to the treatment facility. Coordination is required between the scene and the receiving hospital to ensure that medical personnel are prepared to handle the incoming patients.

Under the circumstances in which very large numbers of individuals need to be decontaminated, it will not be possible to provide individual oversight of each individual. In such situations, individuals that are expected to be contaminated should be transported to suitable locations (*e.g.*, sport centers, military installations) where large shower facilities are available and/or, in good weather conditions, to temporary outdoor facilities organized to accommodate this procedure. In some cases, the authorities might consider issuing guidance for people to shower in their own homes while waiting to be evacuated.

4.4 Hospital Management of Radiation Casualties

There are a number of useful texts that describe in detail the treatment of patients who have received large absorbed doses of ionizing radiation (AFRRI, 1999; Browne *et al.*, 1989; MacVittie *et al.*, 1996; Mettler and Upton, 1995). Current medical advice for the treatment of radiation casualties can be obtained from two organizations within the United States. The first is the Radiation Emergency Assistance Center/Training Site, sponsored by the U.S. Department of Energy (DOE), and managed under the Oak Ridge Institute for Science and Education. The second is the Medical Radiobiology Advisory Team, sponsored by the U.S. Department of Defense (DOD) and managed by AFRRI. Contact information for these two organizations is provided below.

Radiation Emergency Assistance Center/Training Site (REAC/TS)
Oak Ridge Institute for Science and Education
PO Box 117, MS 39, Oak Ridge, TN 37831-0117
(865) 576-1005

Medical Radiobiology Advisory Team (MRAT)
Armed Forces Radiobiology Research Institute
National Naval Medical Center
8901 Wisconsin Avenue, Building 42
Bethesda, MD 20889-5603
(301) 295-0316

TABLE 4.2—*Recommended procedures for on-scene responders.*[a]

1. On-scene responders should wear gloves and a gown or other protective clothing. Each responder should be provided with a personal dosimetry device.
2. Medically unstable patients should be transported to a hospital immediately. A radiological survey, decontamination procedures, or steps taken to contain the contamination may be performed in the ambulance provided these actions do not interfere with more immediate medical requirements of the patient.
3. If the patient is medically stable and conditions at the site permit, limit any further exposure to radiation by moving the patient to an area of low background. The outer clothing of the individual should be removed and the patient should be wrapped in a cloth sheet or blanket to permit handling. The wrapping should be loose to avoid hyperthermia and to allow easy access to the patient by medical personnel.
4. Treat the patient's injuries (*i.e.*, burns, cuts, etc.) sustained in the incident and then, if needed, provide symptomatic treatment for the radiation illness (*e.g.*, the use of anti-emetics). If an open wound is involved, cover the wound with a clean dressing.
5. Do not release a medically stable patient to ambulance personnel before a radiological survey has been performed. If contamination is confirmed, a *preliminary* decontamination should be performed. Record the results of the radiological survey and proceed to decontaminate the patient.
6. Decontaminate the medically stable patient by washing the individual with tepid water to remove any radioactive contamination, beginning with the areas of highest levels of contamination. Proceed gently, mindful that this is a preliminary decontamination and that a more thorough decontamination process will be performed at a medical facility. When finished, repeat the radiation survey of the patient and record the final results. Save all clothing and bedding and all metal objects (*e.g.*, jewelry, coins, belt buckles, etc.). A nasal swab is also recommended to detect inhalation of radioactive contaminants. Tag each item with the patient's name, location, time and date. Save each in appropriate containers; mark containers clearly with: "RADIOACTIVE—DO NOT DISCARD."
7. Transport patient to a medical facility for further treatment. The medical facility should be given advanced warning if they are going to receive patients exposed to radiation so that the facility can institute the appropriate medical protocols. Remember, individuals suffering from radiation injury may not be radioactive, but their skin and the clothing they are wearing could be contaminated with radioactive material. Protection of first responders should be focused on the source of the radiation.

While meeting their responsibilities, on-scene responders should be mindful to follow the basic radiation protection principles:

TIME:	Reduce the amount of time exposed.
DISTANCE:	Increase your distance from the radioactive source.
SHIELDING:	Use shielding between you and the source.

[a]Adapted from the 1998 FBI Contingency Plan for Weapons of Mass Destruction (FBI, 1998).

4.4.1 *Hospital Preparations*

Once the hospital emergency room receives notification of the incident, it should immediately initiate its radiological response plan. Since the entrance used for contaminated patients may not be the usual emergency room entrance, the ambulance personnel must be so informed. It is useful for security personnel to be stationed at appropriate locations to provide directions for ambulance drivers and to limit access only to essential personnel. When the ambulance arrives, the patients are conducted into the treatment area. If there is not an outside door to the treatment room, there are several ways to move the patients without spreading contamination. One way is to lay nonskid plastic sheeting down the hallways over which the ambulance stretcher may be wheeled. It is also possible, if the patient's injuries are not too serious, to transfer the patient from the potentially contaminated stretcher in the ambulance onto a clean stretcher with the patient wrapped in clean blankets or sheets. The patient can then be transported down the usual hallways with the contamination contained inside the wrapping.

It must be noted emphatically that radioactive contamination (whether internal or external) is never immediately life threatening and therefore, a radiological assessment or decontamination should never take precedence over significant medical conditions. The general objectives in approximate order of importance for the management of contaminated, injured patients are as follows:

1. First aid and resuscitation
2. Medical stabilization
3. Definitive treatment of serious injuries
4. Prevention/minimization of internal contamination
5. Assessment of external contamination and decontamination
6. Treatment of other minor injuries
7. Containment of the contamination to the treatment area and prevention of contamination of other personnel
8. Minimization of external radiation to treatment personnel
9. Assessment of internal contamination
10. Treatment of internal contamination (this could be concurrent with many of the above)
11. Assessment of local radiation injuries/radiation burns
12. Careful long-term follow up of patients with significant whole-body irradiation or internal contamination
13. Careful counseling of patient and family members about expected long term effects and risks

Aggressive surgery such as amputation or extensive exploration should not be undertaken to eliminate radioactive contamination. The surgical damage will generally far exceed any potential decrease in lifetime radiological exposure risk. Surgery to remove highly radioactive fragments may be indicated to avoid large exposures.

The psychological needs of the patient are all too often forgotten in the emergency management of such patients. Certainly, the emergency room is a strange enough environment for most patients. This feeling, coupled with the appearance of frightened medical staff all suited up in gowns, etc. is even more unsettling to the patient who may have an exaggerated fear of radiation. A calm and reassuring attitude is essential for the care of both the patient and the family. Careful discussion with the patient about the early and long-term effects of the radiation can be as important as any other aspect of the treatment. This discussion should include the reassurance that the radiation exposure or contamination of the patient will not necessarily be a hazard to friends or family members. If temporary precautions involving contact with the patient are recommended, these should be discussed.

4.4.2 *Patients with Wounds or Burns*

If there are open wounds and they are free of contamination, they should be covered with a water-proof dressing to prevent cross-contamination. Contaminated wounds may be cleaned by gentle scrubbing with a surgical sponge and irrigation. Debridement for removal of contamination should be carefully considered and excision of wounds is appropriate when surgically reasonable. Radioactive contaminants will be in the wound surfaces and will be removed with the tissue.

Emergency management of burns that are radioactively contaminated is a difficult problem. The immediate instinct of emergency staff is to thoroughly wash such burns to remove the contamination. This should not be done for several reasons. If the thermal burn is extensive, any washing will place the patient in grave danger of hypothermia and hypotension. Even if the thermal burn is localized, scrubbing may remove marginally viable skin and make the burn treatment much more difficult. As there is no circulation in the burned tissue, contaminants will remain in the layers of dead tissue and when properly handled, the patient has very little chance of internal contamination. Usually, gentle rinsing of local burns is all that is initially necessary. The burn is then covered, and over the next few days the exudate will lift out much of the contamination into the dressings. Blisters should be left closed, open blisters irrigated, and treated in accordance with appropriate burn protocols.

4.4.3 *Treatment of Patients Who Have Received Large Absorbed Doses of Ionizing Radiation*

The following discussion has been adapted from the 1999 Handbook on the Medical Management of Radiological Casualties prepared by AFRRI (1999).

The initial symptoms of a large absorbed dose of ionizing radiation begin within hours of the exposure. They include nausea, vomiting, diarrhea, fatigue, weakness, fever and headache. These symptoms generally do not last longer than 24 to 48 h after exposure, but a vague weakness and fatigue can persist for an undetermined length of time. The time of onset, severity, and duration of these signs are dose and dose-rate dependent. They can be used in conjunction with white blood cell differential counts to determine the presence and severity of the acute radiation syndrome.

Both the rate and degree of decrease in blood cells are dose dependent (Table 4.3). A useful rule-of-thumb: if lymphocytes have decreased by 50 percent and are less than $1 \times 10^3\ \mu L^{-1}$ within 24 to 48 h and no other medical conditions that could cause these symptoms are apparent, the patient has received at least a moderate absorbed dose of radiation (Goans *et al.*, 1997). For patients with combined injuries, lymphocytes may be an unreliable indicator. Patients with severe burns and/or trauma to more than one system often develop lymphopenia. Associated injuries (trauma/burn)

TABLE 4.3—*Lymphocyte count in humans at 24 to 48 h after radiation.*

Lymphocyte Count ($10^3\ \mu L^{-1}$)	Absorbed Dose Range (Gy)	Lethality without Medical Treatment (%)
3	0 – 0.25	—
1.2 – 2	1 – 2	<5
0.4 – 1.2	2 – 3.5	<50
0.1 – 1.2	3.5 – 5	50 – 99
0 – 0.1	>5.5	99 – 100

should be assessed by standard procedures, keeping in mind that the signs and symptoms of tissue injuries can mimic and obscure those caused by acute radiation effects. For these reasons, there is a need to develop biological techniques to reliably and rapidly assess both whole-body and partial-body exposure to ionizing radiation in the absorbed dose range of greatest relevance for successful medical intervention (currently 2 to 10 Gy).

The prevention and management of infection is the mainstay of therapy. Antibiotic prophylaxis should only be considered in afebrile patients at the highest risk for infection. These patients have profound neutropenia, less than 100 cells per microliter with an expected duration of greater than 7 d. The degree of neutropenia is the greatest risk factor for developing infection. As the duration of neutropenia increases, the risk of secondary infections also increases. For these reasons, adjuvant therapies such as those discussed below will prove invaluable in the treatment of the severely irradiated person.

Initial care of medical casualties with moderate and severe radiation exposure should probably include early institution of measures to reduce pathogen acquisition, with emphasis on low-microbial-content food, acceptable water supplies, frequent hand washing (or wearing of gloves), and air filtration. During the neutropenic period, prophylactic use of selective gut decontamination agents with antibiotics that suppress aerobes but preserve anaerobes is recommended. The use of sucralfate or prostaglandin analogues may prevent gastric hemorrhage without decreasing gastric activity. If possible, early oral feeding is preferred to intravenous feeding to maintain the immunologic and physiologic integrity of the gut.

As of this date, no therapeutic agents or regimens have been approved by the Food and Drug Administration (FDA) for the specific treatment of hematopoietic injury from ionizing radiation, but some of the agents that could be efficacious have been approved by the FDA for other purposes. It is the responsibility of the licensed medical provider to decide how best to use whatever therapy is available at that time in the best interest of the patient.

Hematopoietic growth factors, such as filgrastim (Neupogen®) (granulocyte colony stimulating factor) and sargramostim (Leukine®) (granulocyte-macrophage colony stimulating factor), are potent stimulators of hematopoiesis and shorten the time to recovery of neutrophils. The risk of infection and subsequent complications are directly related to depth and duration of neutropenia. Clinical support should be in the form of antibiotics and fresh, irradiated platelets and blood products. Used concurrently with filgrastim or sargramostim, a marked reduction in infectious complications reduces morbidity and mortality. The longer the duration of severe neutropenia the greater the risk of secondary infections. An additional benefit of the colony stimulating factors is their ability to increase the functional capacity of the neutrophil and thereby contribute to the prevention of infection as an active part of cellular host defense. In order to achieve maximum clinical response, filgrastim or sargramostim should be started within 24 to 72 h subsequent to the exposure. This provides the opportunity for maximum recovery. Cytokine administration should continue, with daily consecutive injections, to reach the desired effect. Bone marrow or cord stem cell transplants have not been shown to improve survival.

4.4.4 *Treatment of Internal Contamination*

Medical treatment to facilitate excretion and/or reduce incorporation of radionuclides will occur in a hospital after stabilization and decontamination. Specific treatment protocols will be tailored to the particular radionuclides that have been incorporated in the body. Detailed procedures to dilute, purge or facilitate fecal and/or urinary elimination of radionuclides, and thereby reduce the absorbed dose to the person, are discussed in NCRP Report No. 65 (NCRP, 1980b).

Potassium iodide (KI) administration can be used to reduce radiation exposure to the thyroid gland from radioactive iodines. Oral administration of 130 mg of KI at or before exposure to radioactive iodines effectively blocks close to 100 percent of radioiodine from reaching the thyroid. Unfortunately, the effectiveness of this intervention decreases rapidly with time after exposure. Administration of KI 4 h after exposure will only block about 50 percent and administration more than 12 h post-exposure will have little effect. There is some value if the exposure to radioactive iodine is expected to continue over several hours or days.

In most cases, the value of KI administration is expected to be low. Even when there is a significant release of radioactive iodine (*e.g.*, following a nuclear yield from a weapon), the value of administration of KI to adults is small. Recent analysis of epidemiological studies following external radiation exposure of the thyroid, have shown little if any risk to persons exposed over the age of 20 (Ron *et al.*, 1995). For a number of reasons, the carcinogenic effect of radioiodine is felt to be even less than from external radiation. As a result there is little reason to consider large programs to distribute KI to adults in the event of terrorist incidents.

However, experience from a number of epidemiological studies as well as more recent experience from Chernobyl indicates that the thyroid of the fetus and child is likely to be quite sensitive to induction of thyroid cancer following radiation exposure. As a result, there is reason to have a plan to distribute KI to pregnant females and to children if a terrorist scenario is suspected to involve a nuclear weapon or some other major release of radioiodine. Primate studies have shown that administration of KI to pregnant females in the amount listed above will effectively block the fetal thyroid (Noteboom *et al.*, 1997).

4.4.5 *Combined Injuries*

Section 3 outlines the complexities of energy release from a nuclear device where blast and thermal effects may be superimposed on the effects of radiation fallout (residual radiation) as well as the initial radiation that is released within moments of the detonation. With detonation of a nuclear weapon, people may not be killed outright by the blast or thermal effects, but could have burns or wounds from flying debris, and perhaps fractures or other traumatic injuries that are superimposed on whatever radiation injuries they have sustained. Such combined injuries increase greatly the chance of death and worsen the condition of the casualty. Under circumstances where an exposure to radiation, a burn or other injury alone would not be lethal, a burn or other wound superimposed on sublethal radiation injury could lead to infection and rapid death. Whole-body absorbed doses as low as 0.5 Gy may be sufficiently suppressive on the immune system as to predispose the irradiated individual to either bacterial, viral or fungal infections. These types of infections were major problems among firefighters at Chernobyl. Extreme care should be taken to minimize the chances of infection in persons with immune systems compromised by ionizing radiation.

With a detonation of a nuclear weapon near the ground, the early fallout of large particles containing radionuclides contributes significantly to the absorbed doses received within the first few hours after the blast (see Table 3.4). Beta radiation as well as the more penetrating gamma radiation from this early fallout, if deposited anywhere on the body, can produce extreme local damage to skin, muscle, connective tissues, bone, and other tissues. Multiple surgeries could be required to remove necrotic and infected tissues.

Decontamination and debridement can greatly reduce this localized radiation injury. The term "cutaneous syndrome" has been used to describe the localized damage to large skin areas from "beta burns" (Gottlober *et al.*, 1996; Peter *et al.*, 1999).

Other modes of combined injury include inhalation of large or small radioactive particles with the attendant possibility of lung damage, the accumulation of fluids and pneumonia. At large absorbed doses, if the patient survives the initial effects, lung fibrosis may also develop. Late arising neoplasias are also a major concern, especially within exposed individuals with preexisting pulmonary disease (*e.g.*, asthmatics, heavy smokers, etc.) (NAS/NRC, 1988). Lung damage predisposing to pneumonia could be lethal in a person who has been rendered more susceptible to infection by external irradiation that damages the bone marrow and reduces circulating white blood cells. In spite of this, lung lavage to remove radioactive particles is not generally recommended.

A certain class of chemical warfare agents, specifically the mustards, produces extensive skin damage, which could become infected and thus compromise the survival of irradiated people. Some portion of these agents also are transported through the blood to the bone marrow where they kill dividing marrow cells and would add to the damage produced by any radiation injury already sustained.

4.5 Pharmacological Radioprotection

There are two basic pharmacological strategies that will limit the risks associated with a radiological incident. The first is the use of agents that will reduce the anticipated exposure from internal contamination by either blocking the absorption of a nuclide or by more rapidly eliminating the nuclide from the body. Because this strategy has been extensively addressed by the NCRP in Report No. 65 (NCRP, 1980b), the reader is referred to that text.

The second strategy reduces long-term risks by reducing the sensitivity of the exposed individual to the deleterious effects of the radiation. It has been known for many years that a variety of compounds can reduce both the late (specifically the elevated risk of cancer and cellular mutations) and the early effects of ionizing radiation. There are several classes of such compounds with different modes of action. Sulfhydril compounds work primarily by scavenging free radicals which in turn minimizes indirect radiation damage to critical sites within the cell. WR-2721, also known as amifostine or by its trade name Ethyol®, is probably the most effective of these compounds (Capizzi, 1999). Amifostine has been approved by the FDA for use as a protectant of normal tissues of the oral cavity during radiotherapy of patients suffering from head and neck cancers. At high concentrations (*e.g.*, 700 to 900 mg m^{-2}), amifostine can provide significant levels of radioprotection as demonstrated clinically by significant reduction in radiation-induced mucositis, and experimentally in animals by the markedly enhanced survival following supralethal exposures to radiation (*e.g.*, dose reduction factors of ~1.8 are commonly reported).[8] At these elevated drug concentrations, however, amifostine has significant toxic side effects, including nausea, vomiting and hypotension. The latter toxic side effects clearly make this drug, when used at such high concentrations, unacceptable for use as a general protective pretreatment for general populations or even selected personnel at increased risk to radiation exposure. By lowering the concentration of amifostine, the side effects can be controlled and therefore could be considered for use to protect selected emergency responders who are likely to receive exposures well in excess of occupational exposure limits. Because of this, pre-screening high-risk personnel for susceptibility to amifostine-induced toxicity might be considered if the drug were to be deployed. Although amifostine appears to lose most of its cytoprotective effects when delivered at nontoxic, nonperformance decrementing exposures, it appears to retain much of its anti-mutagenic, anti-carcinogenic potential (Grdina *et al.*, 1995).

Except in the specific application discussed above, the FDA has not approved any drugs for protection against the effects of ionizing radiation. Nevertheless, the possibility exists that radiation injury could be reduced or repopulation of key tissues or organs could be enhanced by administering certain pharmacological or nutritional agents before irradiation. The point to be stressed is that animal studies have shown that it is possible to protect against injuries to DNA that result in cell death leading to marrow damage and increased susceptibility to infection and to DNA damage that leads to cancer. Medical planners should assess the projected usefulness, efficacy and availability of medicaments that, based on contemporary information, they judge to have application either before or

[8]In this case, the dose reduction factor is defined to be the ratio of the mean lethal dose with the use of a protective agent divided by the mean lethal dose without the use of the agent.

after irradiation. The essential point is that pharmacological intervention strategies exist and should be considered

4.6 Medical Follow-Up of Persons Exposed to Ionizing Radiation

There are often questions raised about how to provide follow-up for persons exposed to ionizing radiation. These issues extend across a very wide spectrum from quite low absorbed doses received by persons living around normally operating nuclear facilities to very high absorbed doses incurred in radiotherapy or radiation accidents.

The problem must take into account several factors. The first is a definition of the level of risk. This should be based on the absorbed dose to an organ of interest, the radiation type and quality, and if possible, the use of the appropriate age- and tumor-specific risk factor. The risks to a specific individual should not be based on estimates of the effective dose and nominal risk coefficients. If the estimated risk is low relative to spontaneous disease rates, then no follow-up is necessary.

If the risk of a radiogenic disease is determined to be high, then an analysis of the magnitude of the exposure, the age of the patient when the exposure occurred, and other relevant factors is important to take into account temporal factors. For example, if radiogenic lung cancer is the issue, then there would be little reason to do screening before the minimum latent period of 10 y or so. Similarly, since the risk of radiogenic leukemia decreases markedly with time, there is little reason to screen beyond 30 y after exposure.

Given consideration of these temporal factors, if the risk is determined to be high, then one needs to consider whether there is a screening test for that tumor that is accurate and has been shown to be effective in reducing mortality or improving quality of life. At the present time the American Cancer Society indicates that there are effective screening tests for breast, cervical and, perhaps, colon cancer. While leukemia, lung, stomach and other cancers are known to be radiogenic, there is no screening test that has been recommended for widespread use to detect these tumors. For at least some age groups, the thyroid is sensitive to radiogenic tumor induction. To date most of the tumors found in children at Chernobyl have been found by palpation, although some have been found by ultrasound.

Patients who have been treated for a primary malignancy are often followed to see if a second malignancy develops. This is true regardless of whether they have received radiotherapy, chemotherapy, or even surgery. The secondary tumors of concern after radiotherapy will arise from tissue elements in or near the edge of the radiation treatment port, and they may arise up to decades later.

5. Psychosocial Effects of Radiological Terrorist Incidents[9]

In preparing for, or responding to, a terrorist attack involving radioactive materials, it will be essential to recognize the importance of social and psychological issues. A radiological incident can produce profound psychosocial impacts at all levels of society, affecting individuals, families, communities, and the nation as a whole. For example, in the immediate aftermath of an incident, thousands of people who fear possible exposure to radiation may stream into area medical centers to seek assistance. Even well prepared health and human services facilities would face enormous challenges in dealing with such a situation, while less adequately prepared facilities could easily be overwhelmed and rendered ineffective (Becker, 2001b).

Other psychosocial effects may create longer-term problems. At the individual level, significant numbers of people may suffer chronic distress years after an incident, while at the community level, contamination-related stigma as well as conflict related to the cleanup may considerably hinder recovery efforts. At the broadest level, a radiological incident has the potential to produce widespread fear, a heightened sense of vulnerability, loss of trust, and an overall loss of confidence in societal institutions.

It is vital, therefore, that psychosocial considerations be a high-priority component of consequence-management efforts. Without a well-developed psychosocial component, response efforts after a terrorist attack could be "successful" in a technical sense,

[9]Several of the policy recommendations in this Section were presented at the Workshop on the Management of Nuclear Disasters, International collaboration for Disaster Health Crisis Management, National Hospital Tokyo Disaster Medical Center, Tokyo, Japan (Becker, 2001a). In addition, part of the discussion of radiation incident features and effects is based on Becker (2001b).

but a failure in terms of reducing morbidity and maintaining public trust and confidence. The remainder of this Section examines some of the psychosocial aspects of radiological terrorist incidents, and considers the implications of psychosocial issues for emergency and post-emergency planning, training and response.

5.1 Key Psychosocial Features of Radiological Terrorist Incidents

In a sense, radiological terrorist incidents represent a combination of two broad characteristics, each of which is sufficient to produce serious psychosocial consequences. First, these incidents involve toxic hazards. Second, these incidents are deliberate acts caused by human beings.

As Erikson (1994) has noted, people find situations involving radiation and other toxic agents "a good deal more threatening than both natural hazards of even the most dangerous kind and mechanical mishaps of considerable power." Toxic hazards "unnerve human beings in new and special ways."

Why should this be the case? For one thing, situations involving ionizing radiation (or hazardous chemicals) involve risks that are perceived to be *involuntary* and *unfamiliar*. Both of these features are believed to trigger more concern than other sorts of risks (Bennett, 1999; Stokes and Banderet, 1997). Then too, toxic agents are often *invisible*: they are "without substance and cannot be apprehended by the use of the unaided senses, and for that reason they seem especially terrifying" (Erikson, 1994). Frightening historical associations (*e.g.*, Hiroshima, Nagasaki, Chernobyl) and the generally negative images that people tend to associate with nuclear facilities and activities (Slovic *et al.*, 1993) may contribute further to the sense of dread.

Toxic hazards are also viewed as having the potential to cause *hidden and irreversible* damage and as having the capacity to produce *forms of illness and death that arouse particular dread*. In addition, such situations are seen to represent *special dangers to children and pregnant women*. Again, all of these factors are thought to be connected with a greater sense of alarm (Bennett, 1999).

Furthermore, invisible contaminants are seen as posing an unbounded or *open-ended threat*. Because long-term health consequences may take years to develop, the danger is seen as having no end. There is a continuing sense of vulnerability and concern, and people can remain in a "permanent state of alarm and anxiety" (Erikson, 1994). As Ursano *et al.* (1994a) have written, contamination incidents "produce long-term anticipatory stress of the possible, the probable and the imagined risks to health and family."

Also playing a role is the fact that situations involving toxic hazards are a result of human activities; that is, they are human-made in origin. People who are victimized by such events "feel a special measure of distress when they come to think that their affliction was caused by other human beings" (Erikson, 1994).

Taken together, these features and perceptions make radiation a powerful stressor. "The insidious and lethal nature of radiation," writes Mickley (1989), "makes it especially feared." Similarly, Rosa and Freudenburg (1993), drawing on the work of Slovic *et al.* (1993), note that "nuclear risks are perceived to be the riskiest—and are the most dreaded."

If the first broad element of a radiological incident that can produce psychosocial impacts is the fact that invisible toxic agents are involved, the second key element is that the disaster is not a mistake or an accident, but an act of **terrorism**—a calculated, *intentional act* intended to inflict pain, injury, suffering and death. Terrorist attacks, researchers and practitioners have suggested, can produce especially high psychological morbidity. In the words of a recent report by the Institute of Medicine/National Research Council (IOM/NRC, 1999a), "the literature on civilian terrorist attacks reveals a number of reports of very high rates of Post-Traumatic Stress Disorder (PTSD) after such attacks." Furthermore, notes the report, "in addition to PTSD, many of the victims of a terrorist attack may suffer the death of family members, close friends, or work colleagues, which can lead to a complicated bereavement with its own elevated risk for depression, self medication, and substance abuse."

These findings are echoed in observations made after the Oklahoma City bombing. Reflecting on the psychosocial effects of the 1995 terrorist attack, Dr. Brian Flynn, Chief of the Emergency Services and Disaster Relief Branch at the U.S. Public Health Service (PHS), has commented: "As someone who has been responding to the psychological needs of survivors of mostly natural disasters for many years, I am fairly certain that the psychological sequelae we are seeing following the Oklahoma City bombing are more pervasive, more severe, and of longer duration than our natural disaster experience" (Flynn, 1996).

5.2 Spectrum of Psychosocial Effects

The combination of characteristics identified above—invisible, toxic agents coupled with terrorism—means that a radiological attack will have a powerful capacity to produce a wide range of acute and chronic psychosocial impacts. At one end of the spectrum are the common stress reactions that are typically associated with both natural and human-made disasters. Terrorist-initiated exposures to ionizing radiation or release of radioactive materials may produce a variety of such reactions in survivors, emergency responders, and others. The effects can be emotional, physical, cognitive or interpersonal in nature, ranging, for example, from fatigue, insomnia or impaired concentration to emotional numbing or social withdrawal (Young *et al.*, 1998). As is the case with other types of disaster, many mild to moderate stress reactions are transient in nature. Such reactions represent a normal reaction to a highly abnormal situation (Myers, 1994). "Relief from stress and the passage of time usually lead to the reestablishment of equilibrium, but information about normal reactions, education about ways to handle them, and early attention to symptoms can speed recovery and prevent long-term problems" (Hartsough and Myers, 1985).

What complicates the picture considerably, however, is that exposure to invisible contaminants has also been shown to produce a *chronic* state of alarm. Whether the source of the danger is removed from the community, or alternatively, whether the survivors are relocated away from the danger zone, people continue to have serious concerns about the health implications of the incident. In other words, although the immediate threat may be over, and although considerable time may have passed, the incident continues to act as a powerful and persistent stressor.

Research suggests that chronic stress may be a primary psychosocial aftereffect of nuclear reactor accidents. Studies carried out more than 6 y after the Chernobyl disaster found a high prevalence of psychological distress and psychiatric disorders (mainly milder psychiatric syndromes) in the severely contaminated Gomel Region (Havenaar *et al.*, 1996). A comparison of this area with a control region found significantly higher levels of psychopathology among the exposed population (Havenaar *et al.*, 1997). While the effects in the overall population in the exposed area were mainly at a sub-clinical level, a significantly higher risk of psychiatric disorders was found among mothers with children under 18 y of age. The researchers speculate that "psychiatric symptoms among these women are fostered by genuine concern about the health of their children..."

The Chernobyl studies reinforce earlier findings about the effects of the March 28, 1979 accident at the TMI Nuclear Power Plant in Middletown, Pennsylvania. Studies conducted by Bromet *et al.* (1990) found that the accident had a long-term adverse effect on the mental health of mothers of young children. Likewise, research by Baum *et al.* (1983) and his colleagues found evidence of long-term emotional, behavioral and physiological stress after TMI. Likewise, Baum *et al.* (1983), Collins (1991), and other researchers have found evidence of long-term emotional, behavioral and physiological stress after TMI.

Persistent stress can produce a marked deterioration in the quality of life. At a minimum, having long-term health concerns about loved ones as a more or less permanent feature of day-to-day existence very likely affects attitudes toward the present and hopes for the future. Further, it is now widely accepted that long-term stress is deleterious to health (*e.g.*, high blood pressure, stroke, heart disease). In a situation where large numbers of people are suffering from chronic, unremitting stress related to a radiological incident, this could translate into substantial problems of physical illness (Baum and Fleming, 1993) and significantly increased demand on health care facilities.

It is also extremely important to note that chronic stress reactions can develop *even in the absence of actual exposure*. In research carried out after the Goiania radiological accident in Brazil, Collins and de Carvalho (1993) compared stress levels in three groups: people who had been exposed to radiation as a result of the accident, individuals who had *not* been exposed but who lived between 1,500 and 1,700 m from a radioactive waste storage facility, and a control group. One of the most striking findings was that people in both the exposed group *and* the group that was not exposed but was concerned about living near the waste facility showed psychological, behavioral and cardiovascular-neuroendocrine effects. For example, both groups reported more fear than controls, both exhibited decrements in performance on speed and accuracy tests, and both had significantly higher blood pressure than controls. Equally striking was the fact that these effects were present 3.5 y after the accident. The authors concluded: "Anticipatory stress associated with potential exposure to ionizing radiation resulted in a level of stress similar to that from actual exposure to ionizing radiation." Clearly these findings further reinforce the idea that psychosocial casualties after a radiological terrorist incident could extend far beyond the immediate area of impact.

People involved in a radiological incident may also experience other effects, including survivor guilt or a feeling of somehow being responsible for loved ones being exposed, a sense of betrayal, or a loss of trust. As in other disasters, survivors of a radiological incident may also have to deal with a new sense of vulnerability. In a disaster, "the fabric of everyday existence is torn away to reveal danger and risk ... Once something of this nature has happened to a person, it is very difficult ... to believe life can ever be the same again..." (Hodgkinson and Stewart, 1998).

Meanwhile, for a portion of the population, there is the risk that serious and persistent mental health problems may develop or be exacerbated as a result of exposure to the extreme trauma of a radiological incident. Problems can include depression, anxiety disorders, substance abuse, and PTSD.

PTSD is "a prolonged post-traumatic stress response" (APA, 1994; Young *et al.*, 1998). Among the features associated with PTSD are a persistent reexperiencing of a traumatic event, persistent avoidance of stimuli associated with the trauma and a numbing of general responsiveness, and persistent symptoms of increased arousal (*e.g.*, irritability or outbursts of anger, exaggerated startle response). For a diagnosis of PTSD to be made, the disturbance must last for at least one month, and there must be "clinically significant distress or impairment in social, occupational, or other important areas of functioning" (APA, 1994).

According to Ursano *et al.* (1994b), "one of the best predictors of psychiatric illness after a traumatic event is the severity of the trauma." The greatest risk of PTSD is in persons "exposed to life threats and perhaps, in those exposed to terror, horror, the grotesque..." (Fullerton and Ursano, 1997a). To the extent that a radiological incident has such characteristics, it can reasonably be expected to have the potential to cause more serious problems. A scenario in which survivors have witnessed horrific radiation burns or similar injuries would clearly fall into this category. Mickley (1989), drawing on the well-known research of Lifton (1967) on Hiroshima, comments that "seeing large numbers of burned, cut, and maimed bodies was a major source of emotional trauma after the bombing of Hiroshima." Incidents that totally destroy a community's support structure can also reasonably be expected to produce more psychosocial casualties, since social support is a critically important factor in people's ability to cope with trauma and adversity.

In sum, the combination of invisible hazardous agents and the intentionality and horror of the terrorist act may combine to make a radiological terrorist incident an extremely traumatic event with the potential to cause substantial psychosocial harm.

5.3 Identifying and Assisting High-Risk Groups

In the literature on disaster, several sub-populations are usually identified as being at greater risk of psychosocial impacts or as having special needs.

5.3.1 *Children*

In the aftermath of a radiological attack, the psychosocial well-being of children will need to be another focus of assistance efforts. Children are "a particularly vulnerable group and require special attention and programs" (Farberow and Gordon, 1981). Appropriate mental health services and psychosocial support need to be available to help children and adolescents (and their families), just as assistance is provided to other groups at high risk for traumatic exposures. Such assistance efforts should be informed by a current understanding of children's post-disaster reactions, relevant techniques for the assessment of children, and strategies of intervention (Pynoos *et al.*, 1998).

5.3.2 *Emergency Workers / Responders*

Another group that always warrants special attention is emergency/disaster workers. These personnel can encounter extraordinary stresses and highly traumatic situations in the line of duty. This may include facing threats to their own health and safety, having to handle mutilated bodies, etc. (Ursano *et al.*, 1994a). This can put such workers at higher risk for psychological effects. Special attention, therefore, needs to be devoted to the well being of responders and their families (Fullerton and Ursano, 1997b).

The threat of exposure that is associated with operations in a potentially contaminated environment is likely to add considerably to the stresses faced by emergency workers. Conducting urban search and rescue and other operations in the aftermath of an explosion such as the one involving the bombing of the Oklahoma City Federal Building in 1995 is already quite taxing; were such an incident also to involve significant danger from radiation, stresses could be greatly magnified. For one thing, there is the increased threat to personal safety. In addition, there may be cases where emergency workers are aware of someone who needs assistance, but are ordered not to go in due to potentially lethal levels of ionizing radiation. Such a situation could be stressful in the extreme. Finally, having to work in personal protective equipment can bring its own added stresses, underlining the importance of frequent and extensive training (Carter and Cammermeyer, 1985; Fullerton and Ursano, 1994; Stokes and Banderet, 1997).

5.3.3 *Pregnant Women and Mothers with Young Children*

Another high-risk group warranting special attention in the aftermath of a radiological incident is composed of pregnant women and mothers with young children. As noted earlier, studies carried out after both the accidents at TMI and Chernobyl identified women with young children as being at significantly greater risk for psychological effects than the general population. While there will be some differences between an accident and a radiological terrorist incident, it is likely that the same kinds of concerns about radioactivity and its future effects on the health of children will come into play. Pregnant women face the same concerns but also may feel additional pressure to consider an abortion to preclude giving birth to a malformed child. Even at fetal absorbed doses at which adverse effects are unlikely, the additional anxiety associated with such a decision may be significant. In light of these issues, health and human service providers will need to be prepared to implement special services and interventions.

5.3.4 *Other High-Risk Groups*

Cleanup workers have been identified as being at higher risk of psychological effects. People who were involved in cleaning up the Chernobyl nuclear accident, for example, have been reported to be at greater risk for a range of social and psychological problems (Koscheyev *et al.*, 1997). The Chernobyl experience also suggests that *evacuees* may be at greater risk (Havenaar *et al.*, 1996). Other high-risk groups include *older people*, since they may have limited support networks, mobility impairment, illnesses, etc. (Young *et al.*, 1998), and *people with mental illness* or psychiatric disabilities (DHHS, 1996).

5.4 Potential Magnitude of Psychosocial Impacts After a Radiological Terrorist Incident

Situations involving the release of invisible hazardous agents have the potential to produce psychosocial effects in a very large number of people. The 1995 sarin attack in the Tokyo subway system, which has been widely discussed in the domestic preparedness community, provides a dramatic illustration of this point. Following the attack, in which 12 individuals were killed, over 5,000 people sought care from area hospitals. The vast majority did not wait for ambulances or other emergency vehicles; instead, they went directly to the nearest medical facility. As has been widely reported, one of the most common problems was acute anxiety and stress related to the attack (Lillibridge and Sidell, 1995). The 1987 accident in Goiania, Brazil provides a similar illustration. In the aftermath of the radiological contamination incident, over 112,000 people sought medical examinations (Collins and de Carvalho, 1993; IAEA, 1988a). According to de Carvalho (1999), "the fear was so intense that some people fainted in the queues as they approached their moment of monitoring. Many complained of vomiting and diarrhea…"

One lesson is clear: It is highly likely that after a radiological incident, extraordinary demands will be placed on the health and human service system. Facilities may be deluged with very large numbers of people seeking help. Regardless of what symptoms those people will be experiencing—from radiation injuries to acute stress—they will all require examinations and care if longer-term problems are to be avoided. Even a well-prepared health and human service system could easily be overwhelmed. In such a situation, survivors might not receive needed care and suffering might be prolonged. In addition, such a situation could easily damage morale and result in a serious loss of public confidence.

With such high numbers of people suffering psychosocial effects, it is essential that psychosocial services be well integrated into the overall health response. Just as local, state and federal medical assets need to receive information on actual or suspected terrorist incidents as early as possible so as to be able to prepare and mobilize effectively, so too do professionals who specialize in social and psychosocial issues need to be brought in at an early stage.

Given the large numbers of people who may transport themselves to hospitals and other emergency facilities, it will also be important to have effective triage protocols as well as screening methods for differentiating the more serious psychosocial problems from less serious ones (IOM/NRC, 1999a). Furthermore, it will be crucial to be prepared for a range of scenarios and incident types, including one where the effects are almost exclusively psychosocial in nature. One example would be an incident in which a very low level of radiological contamination is spread over a densely populated area, accompanied by frightening statements from a terrorist group. There would be no immediate casualties in the conventional sense, but psychosocial effects could be extremely widespread and potentially very long lasting.

5.5 The Problem of Social Stigma

One of the most troubling and persistent impacts of incidents involving radiation is the problem of social stigma. Residents of affected communities may be seen by others as "tainted" and as "people to be avoided" (Edelstein, 1988; Kroll-Smith and Couch, 1993). Social stigma can be powerful and pervasive. Following the radiological contamination at Goiania, Brazil, people from the city found themselves the focus of fears and the targets of discrimination. As Kasperson and Kasperson (1996) have noted: "Hotels in other parts of Brazil refused to allow Goiania residents to register. Some airline pilots refused to fly airplanes that had Goiania residents aboard. Cars with Goiania license plates were stoned in other parts of Brazil."

Most "conventional" disaster recovery and rehabilitation efforts have not had to address or cope with stigma as a major *defining feature* of the situation, and local, state and federal agencies have not devoted a great deal of attention to such issues. Because there is a strong likelihood that stigma will be a significant problem in the aftermath of a radiological incident, and because it could hamper recovery efforts, it is important that officials have in place a plan for dealing with it. Such a plan should be informed by current social and behavioral science research on stigma after contamination episodes, and should include a multidimensional approach that incorporates education programs, media campaigns, high-profile visits by public figures, community forums, and other measures.

5.6 Prevention as the Guiding Principle

It is worth emphasizing that the guiding principle in relation to psychosocial issues should be *prevention*. It is far more effective to intervene early to prevent social and psychological problems from developing than it is to have to address serious problems once they have arisen. What this implies is the need to have plans, infrastructure, resources and trained personnel already in place. In other words, the social and psychosocial component cannot be an afterthought. The time to familiarize health and human services professionals with the dynamics of incidents involving radiation is not after an incident has taken place. The cost of inadequate preparedness is greater morbidity and more long-term effects.

Part of prevention also means trying to make sure that response operations themselves do not exacerbate psychosocial impacts or inflict additional psychosocial injury. For example, the experience of having to disrobe in a relatively public place among large numbers of people (and possibly the media) as part of mass decontamination procedures can be a traumatic experience. More generally, emergency measures and actions, and later consequence-management decisions, need to be informed by an awareness of potential psychosocial implications.

5.7 A Shortage of Resources, Training, and Trained Personnel

To achieve a satisfactory level of preparedness for addressing the psychosocial consequences of the radiological terrorist threat will require experienced personnel, high quality training, and some specialized resources. Unfortunately, at the present time, relatively few medical and human services professionals have had training or experience related to radiological incidents.

Without a thorough knowledge of the dynamics and distinctive characteristics of radiation-related incidents, the effectiveness of helping professionals could be significantly reduced. Furthermore, while working in any disaster environment can be highly stressful, weapons of mass destruction (WMD) incidents bring with them their own additional sets of worries and challenges. Mental health professionals are not immune from preconceptions or fears of their own, and these are likely to be greater in the absence of training and experience.

Despite the obvious need for training in this area, few resources focused specifically on the psychosocial aspects of radiological incidents are presently available. For example, despite its many strengths, the 120 cities Domestic Preparedness Program[10]

[10]Established under Title XIV of Public Law 104-201, the National Defense Authorization Act for Fiscal Year 1997, also known as Nunn-Lugar-Domenici legislation.

includes only a very limited introduction to psychosocial issues in WMD situations. The same is true of other WMD training courses. And at the present time only a very small number of universities in the United States incorporate substantial, up-to-date coverage of psychosocial issues in their courses on chemical and radiological incident management. Likewise, very few institutions offer medical, health and human services continuing education programs focused specifically on the psychosocial aspects of WMD incidents. Clearly, there is a pressing need for additional training opportunities (Becker, 2000).

5.8 Training Exercises

With few exceptions, the radiological incident training exercises conducted to date have not included a substantial psychosocial component. This is true with respect to exercises involving lost sources, reactor accidents, or WMD incidents. Even where psychosocial considerations have been taken into account to some degree, it has generally been only acute or emergency phase impacts that have been addressed. More often than not, social and behavioral issues have been dealt with "on paper" or as "assumptions" in radiological training exercises. Rarely have such issues been worked through and practiced as part of the exercise.

The lack of a substantial psychosocial component in radiological training exercises means that agencies and responders are losing a vital opportunity to become familiar with a crucial element of consequence management. Also lost is the chance to improve and enhance psychosocial response efforts, and to smoothly integrate such efforts into the overall response. Finally, training scenarios may be utilizing unrealistic assumptions about human behavior in contamination situations. Taken as a whole, these problems translate into a reduced capacity to respond effectively to the effects of radiological incidents.

5.9 Research

Studies conducted after TMI, Goiania, Chernobyl, and other accidents have enormously expanded our understanding of the social and behavioral impacts of radiological releases. Much of this information may be relevant to psychosocial consequence-management efforts after a radiological incident. Nevertheless, many gaps in our knowledge remain, and much research remains to be done.

One key area involves psychosocial interventions. Considerable experience has been gained in recent decades from psychosocial interventions undertaken after radiological accidents. However, relatively little in the way of systematic analysis of such efforts has been conducted; indeed, some cases have not even been documented. Because knowledge from these accident interventions may provide valuable insights relevant to service delivery after a radiological terrorist incident, it is important that such interventions be recorded, studied, analyzed and evaluated. Without careful evaluation studies, it is difficult to know with any degree of certainty "what works" and what does not.

In a related vein, more work needs to be done to assess the value of psychological debriefing. This approach is widely used with emergency workers, and many people have reported finding the technique useful (Everly and Mitchell, 1999; Mitchell and Everly, 1996). Research findings to-date, however, have been mixed (Young *et al.*, 1998). Clearly, additional studies are needed, both of the technique in general and of its applicability in contamination situations.

Further research is also needed to identify high-risk groups and to develop appropriate information, support and assistance programs. What measures can be taken to prevent or reduce psychosocial effects in emergency workers, evacuees, cleanup workers and other groups that have already been identified as being at increased risk for psychological effects? How are children affected by an incident involving a WMD, and what special measures and/or interventions can help protect children in such situations? What interventions are most effective in preventing and/or mitigating psychosocial effects among mothers with young children?

Additional research is also needed on people's reactions to decontamination situations, including the psychological effects of undergoing decontamination and ways of reducing the impact of such situations (IOM/NRC, 1999b). Likewise, the problem of stigma requires additional attention from the research community. In particular, there is a pressing need to develop a better understanding of ways of preventing or ameliorating stigma in contamination situations. Finally, ethical issues associated with long-term medical and psychosocial monitoring should also be further explored.

5.10 Restoring and Maintaining Trust

In considering the psychosocial consequences of a radiological incident, the restoration and maintenance of trust will be a key issue. Experience with nuclear reactor accidents suggests that such events have a powerful capacity to produce anger, outrage, resentment and mistrust. Erikson's (1994) comments are germane: "Technological disasters ... being of human manufacture, are at least in principle preventable, so there is always a story to be told about them, always a moral to be drawn from them, always a share of blame to be assigned."

There is no reason to suppose that anger and blame won't be issues in the aftermath of a terrorist-related incident involving radiological materials. Undoubtedly, the initial focus for these emotions will be on the perpetrators of the act (if they can be identified). Yet, depending on how events unfold, it is also possible that with the passage of time, anger could be redirected elsewhere. In some situations, members of the public could feel "let down" by officials who "should have protected them," just as after TMI and other technological disasters many people felt that they had been let down or misled by safety officials. Likewise, if post-incident follow-up is inadequate, if health and human service delivery is badly handled, or if other key aspects of post-disaster programs are not well conducted, people could experience a loss of confidence and redirect some or all of their anger toward government.

To maintain trust and public confidence, decisions that affect the community will need to be grounded in an approach that is open. "Trust is generally reinforced by 'openness,' not only in the sense of making information available, but in giving a candid account of the evidence underlying decisions and how it was used" (Bennett, 1999). Difficult and sometimes controversial issues will undoubtedly arise, such as what to do with contaminated debris and soil. These kinds of decisions are likely to engender many of the same kinds of concerns that siting decisions do under normal conditions. In a post-disaster environment, it will be even more crucial for decision making to be transparent. It will also be important to ensure that such decisions don't further stigmatize affected areas, or end up turning communities against each other.

Decision-making processes also need to be participatory. One approach to achieving this is to employ stakeholder advisory boards that actively involve a wide range of participants. For example, community recovery decisions might be guided by a board composed of many sectors of the community, including health and human services professionals, business, media, scientists, representatives of citizen groups, the Red Cross, etc.

Another part of restoring trust will be having mechanisms in place for long-term health care follow-up. Just how such follow-up should be done will require a great deal of careful consideration with respect to such issues as how to avoid the further stigmatization of affected people, how to involve communities in environmental health studies (Stockwell and Smith, 1998), how best to inform people of the risk of illness, and how to maintain privacy and confidentiality.

5.11 Conclusion: The Centrality of the Psychosocial Dimension in Consequence Management

Psychosocial effects are likely to be a critically important aspect of any radiological terrorism incident. Consideration of psychosocial factors, therefore, needs to be an integral part of domestic preparedness efforts, including planning, education, training, research, program development, and response operations. To achieve this end will require that psychosocial issues receive much greater emphasis in the next few years than they have to date.

It will also be important to recognize that psychosocial effects may constitute some of the most challenging aspects of consequence management after a radiological incident (Becker, 1997; Ricks and Berger, 1991). The large number of people experiencing psychosocial effects could easily overwhelm an inadequately prepared health and human services system. In addition, in the aftermath of an incident, anger and mistrust could affect the recovery environment. Likewise, social stigma could result in a sense of isolation among survivors, further complicating assistance efforts.

The chronic nature of some of the psychosocial effects of exposure to invisible contaminants also creates important new challenges. Most current response planning, and most existing programs, are focused on the emergency phase or the weeks and months that follow. With some of the psychosocial impacts of a radiological incident having the potential to be long-lived, it will be vital to focus additional attention on post-emergency response issues. In this regard, it will be important to consider how best to organize long-term medical and psychosocial follow-up into a fully integrated whole. Further research on ways of ameliorating chronic social and psychological impacts will need to be conducted, as will additional evaluation studies of existing approaches.

The issue of psychosocial effects after a radiological (or other WMD) incident is still very much in development. Many questions remain and much research still needs to be conducted. At present, there is no standard protocol or comprehensive manual available. However, a range of steps can be taken to enhance our capabilities for addressing such an incident and reducing its consequences. The following recommendations may be of particular interest to responders, health and human services providers, program developers, planners and policy makers.

5.12 Recommendations

Few emergency plans at the local, state or federal level currently include a well-developed component related to the psychosocial effects of a radiological terrorist incident. Given the large numbers of people that may be affected, and the potential repercussions in terms of confidence in societal institutions, higher priority needs to be given to psychosocial issues in WMD response plans.

Emergency plans for hospitals and other medical facilities should include provisions for dealing with the large number of people who may self-report after a radiological incident. Because many of those who arrive are likely to be experiencing psychosomatic symptoms that mimic symptoms of radiation exposure (*e.g.*, nausea, vomiting, rashes), it will be important both to have effective triage procedures in place, and to be able to recognize and manage large numbers of psychological casualties. In addition, emergency plans should anticipate large numbers of concerned family members, members of the press, and interested members of the public appearing at the hospital, telephoning with questions, etc. Finally, hospital and other emergency plans should ensure that adequate support is available for health care personnel, since the stress of working long hours in a contamination situation may take a toll on them as well.

Programs to train hospital personnel to deal with a large-scale radiological incidents should endeavor to provide information not only to doctors and nurses, but also to other personnel. For example, ancillary staff and support personnel are vital to overall operations, yet they rarely receive any training or information. There is "little reason to believe that medical personnel (including ancillary staff, *e.g.*, housekeepers, central supply workers), inexperienced and perhaps untrained" in WMD incidents, "will be spared from the anxiety and other psychological stresses that will afflict the rest of the community, particularly if the offending agent threatens their own families" (DiGiovanni, 1999).

Although hospitals and health care facilities generally have behavioral health staff on-site or on-call, and although some behavioral staff members may even have had some training in disaster or mass casualty situations, only a very small minority of hospital social workers, psychiatrists, counselors, psychologists and psychiatric nurses have had training specifically related to the psychosocial effects of large-scale radiological terrorism and other forms of WMD terrorism (IOM/NRC, 1999b). Given the importance of psychosocial factors in such an incident, appropriate training modules on the unique social-behavioral challenges of radiological terrorist incidents need to be developed for hospital-based behavioral health staff. In addition, the availability of such modules needs to be publicized through professional societies and professional publications. Finally, behavioral health and social services staff should be included when WMD training is conducted at a medical facility.

Training exercises are a crucial means of improving preparedness. But training exercises are only useful to the extent that they are similar to the conditions likely to be faced by responders. In order to enhance the robustness and realism of exercises, additional content related to psychosocial issues needs to be incorporated. For example, decontamination or medical management exercises would benefit from having to deal not only with mock physical casualties, but with a *mix* of mock physical and mock psychological casualties—some with injuries, some with radiation exposure, some contaminated, and many others with acute stress-related symptoms (nausea, rashes). Having the roles of concerned family members, worried members of the public, interested members of the press, bystanders, etc. covered by exercise volunteers/participants, and having a flood of mock telephone calls to health care facilities and emergency management officials, could further enhance the degree of realism.

Information that is as accurate and complete as possible should be provided as early as possible to people following an incident. Research suggests that an early lack of accurate information can contribute to both anger and fear (Bowler *et al.*, 1994). Attempts to limit or suppress information only serve to undermine trust and destroy public confidence.

When people are instructed to shelter in place, authorities should endeavor to provide a steady stream of information and updates. If people feel that they don't understand what is happening, or if they perceive the situation as open-ended, their level of stress will be greatly increased.

While a radiological incident will create unique challenges because it involves contamination and terrorism, fundamental principles of disaster mental health should still be borne in mind. These include the following:

1. Prosocial behavior (people helping each other) is far more common in disaster situations than panic, looting, etc. (Tierney, 1993)
2. While stress reactions can affect large numbers of people, most stress reactions after disaster tend to be transient. Programs aimed at providing "information about normal reactions, education about ways to handle them, and early attention to symptoms" represent an important opportunity for health and human services agencies to help "speed recovery and prevent long-term problems" (Hartsough and Myers, 1985). Information about the importance of rest and adequate nutrition can be valuable as well.
3. Psychosocial assistance services such as crisis counseling "are primarily directed toward 'normal' people responding normally to an abnormal situation, and to identifying people who are at risk for severe psychological impairment due to the shock of the disaster" (Young *et al.*, 1998).
4. Wherever possible, efforts should be made to limit or reduce exposure to trauma, death, gruesome scenes, etc. (O'Brien, 1998).
5. Wherever possible, disaster workers and other assistance personnel should be well briefed on what to expect in the affected area. Photos and other information can help people to be better prepared for what they are about to experience (O'Brien, 1998).
6. People do not generally seek out mental health services, making outreach vital. This is especially the case when contamination-related stigma is involved.
7. Efforts to restore social support are crucial. As Figley (1986) notes, "the family, plus the social support system in general, is the single most important resource to emotional recovery from catastrophe."
8. Cultural and other differences need to be carefully considered in the design and implementation of assistance efforts and services. This makes it important to involve stakeholders in planning, program development, and evaluation.

9. Special attention needs to be paid to high-risk groups such as children, disaster workers, etc.

Terms such as "radiophobia" should be avoided when discussing people's concerns about radiation. Aside from the fact that the clinical value of such terms is debatable, words like "radiophobia" could easily be seen as dismissive of people's health concerns or as suggesting that people are behaving in a manner that is somehow "irrational."

Epidemiological follow-up after a radiological terrorist incident should include a well-developed psychosocial component. This will help in obtaining a picture of overall impacts, and may also assist in the identification of high-risk groups.

The medical and psychosocial components of the response effort should be well integrated, both in terms of approach and personnel. A medical response that lacks an adequate psychosocial component may leave important impacts of an incident unaddressed, while a mental health component that is totally divorced from the medical response is likely to be unsuccessful.

Information centers, toll-free hotlines, etc. should also endeavor to use an integrated approach. It might be useful, for example, to have a single point of contact for people to telephone or visit, and to have a multi-profession team on duty. Such a team might include information officers, health physicists, doctors and nurses, mental health staff, etc.

Authorities should ensure that people who fear that they have been exposed to radiation or radioactive contaminants should be given requested preliminary medical examinations as soon as possible. People should never be sent away or "dismissed"; rather, individuals should be treated with respect, and their symptoms—regardless of cause—should be taken seriously.

In cases where people have been exposed to radiation, "the best method available for preventing the development of … adverse psychological effects" is to "provide exposed patients with health care that will enable them to maintain a sense of control over their health" (Vyner, 1988). Doctors and patients will need to collaborate in matching vigilance programs to patient needs. Among other things, strategies for reducing overall risk through lifestyle change may be useful.

There is a need for specialized materials for children to be developed. There are few materials (*e.g.*, coloring books or other items for children) focused specifically on radiological hazards.

It is possible that a radiological incident could impact patterns of reproductive behavior. For example, there could be an increase in legally induced abortions, even in locations that are removed from areas most affected by the release (Knudsen, 1991). It will be important, therefore, to have accurate information and counseling services available to assist people who are making reproductive decisions in the aftermath of an incident.

The maintenance of trust is a crucial component in consequence management. To maintain trust and public confidence, decisions that affect the community will need to be grounded in an approach that is open, participatory and inclusive. Likewise, service development, delivery and evaluation will need to involve key stakeholders through such mechanisms as advisory boards.

The involvement of social and behavioral health specialists in consequence-management efforts should not be seen as a surrogate for public involvement in decision-making processes, nor should "insights into risk perception, fear of cancer, or risk communication be used as a tool simply to assuage public fears" (Wandersman and Hallman, 1993).

There is a pressing need for additional research on the social-behavioral aspects of radiological incidents. Among the key areas requiring attention are the following:

1. Case studies, program analyses and evaluation studies of psychosocial interventions after radiological incidents.
2. Research on people's reactions to decontamination situations, the psychological effects of having to undergo decontamination (IOM/NRC, 1999b), and ways of reducing the overall impact of such situations.
3. Further studies to identify high-risk groups and to develop appropriate interventions.
4. Research on the problem of stigma, and on ways of preventing or ameliorating stigma in contamination situations.

6. Command and Control

Responsible officials will have to make a number of difficult decisions, including how best to control and resolve the terrorist incident, how to assure capture and successful prosecution of the terrorists and their accomplices, how to protect the responding forces, how to deal with the casualties, how to protect the public health and welfare, how to communicate with the public and interact with the media, how to deal with the psychosocial issues, how to manage recovery and remediation, and how to minimize the economic impact of the incident. These management requirements will exist in a complex milieu of political and jurisdictional authorities and will involve the interests of a wide range of stakeholders. Detailed plans to deal with these requirements have been developed and exercised by local, state and federal authorities. While the fundamental principles underlying these plans have not changed over the years, the plans themselves are dynamic, changing documents. Accordingly, the current plans are discussed in some detail in Appendix B. This Section will focus on the structural and organizational challenges that have an impact on the success of the command element in the management of any major disaster regardless of the specific system that is in place.

6.1 Critical Elements of Command and Control

The fundamentals of effective command and control are relatively straightforward. One individual must be in charge, and all responding persons and organizations must recognize the authority of that individual. The response forces should be organized in a manner that assures that each essential function is performed and that all of the available expertise is brought to bear on the problem. Within each responding organization and each functional area, there must be clear lines of authority and responsibility. Finally, processes must be in place to assure that necessary information flows to each element of the responding force and that responsible officials have input into the decision-making process. These underlying principles seem simple but there are significant complicating factors.

6.2 The Federal System

The U.S. Constitution established a federal structure in which powers not specifically assigned to the national government were retained by the states.[11] Responsibility for public health, welfare and safety in the context of response to accidents or incidents of the sort discussed in this Report was not given to the federal government, so that responsibility continues to rest with local and state authorities.[12] In contrast, certain powers were assigned to the federal government, including the responsibility to provide for the national defense and to insure domestic tranquility.[13] In the exercise of these powers, Congress has determined that terrorism involving WMD presents a threat that warrants national level response. Such terrorism is a federal crime.[14] By Executive Order, the Attorney General is the Chief Law Enforcement Officer.[15] By Presidential Decision Directive (PDD, 1995/1998), the Attorney General is charged with responsibility for response to terrorism; this authority has been delegated to the FBI (1998). Although many federal and state agencies play critical roles, it follows that the FBI is in charge of crisis management and that local and state authorities are in charge of consequence management.

[11]U.S. Constitution, Amendment X, provides: "The powers not delegated to the United States by the Constitution, nor prohibited by it to the States, are reserved to the States respectively, or to the people."

[12]The federal government does exercise safety related regulatory power (*e.g.*, the Nuclear Regulatory Commission), and performs health related services (*e.g.*, the U.S. Department of Health and Human Services), pursuant to the Commerce Clause and other provisions of the Constitution. However, Congress has never attempted to extend federal primacy to accident/incident consequence management.

[13]U.S. Constitution, Article IV, Section 4.

[14]U.S. Code, Title 18, Section 2332a.

[15]Executive Order 11396.

6.3 The Fog between Crisis and Consequence Management

While the demarcation between crisis management and consequence management is theoretically clear, there are several potential overlaps. When those in charge of determining courses of action to neutralize the terrorists consider options with potentially significant collateral effects, is that crisis management or consequence management? When events occur without warning and there is still a law enforcement requirement, is such after-the-fact law enforcement activity crisis management? Perhaps the most confounding situation occurs when there are significant consequences before the crisis is resolved. A good example of this possible confounding factor is when one site has experienced a dispersal of radioactive material while one or more other sites are still under the control of the terrorists.

Another complicating factor is our system of jurisdictional authorities. While organizations other than the lead agency clearly must be in a supporting role, each agency will respond pursuant to its own statutory responsibilities and authorities. The U.S. Environmental Protection Agency (EPA), for example, currently has independent statutory authority to respond to threats to humans and natural resources. Similar statutory authorities exist at local and state levels. These authorities are going to respond to the incident without seeking or needing the permission of the lead agency. That greatly complicates the management challenge of maintaining effective command and control over the incident.

Finally, the lead agency may well change over time. When responding to a threat, the FBI unquestionably has the lead and will manage the overall response. If one or more sites begin to experience consequences, the consequence-management elements may assume the lead with regard to specific portions of the response. As time passes, there may be yet another shift in primary responsibility [*e.g.*, from the Federal Emergency Management Agency (FEMA) to EPA].

There is no institutional or programmatic method to avoid the types of conflicts described above. Situations in which there are clearly overlapping roles and responsibilities can best be handled by a cooperation motivated from the common goal of the welfare of the public. Conducting exercises and drills, discussed in Section 11, can help to build teamwork and interagency confidence that will help to minimize these difficulties in real situations.

6.4 Command and Control Plans

Local and state authorities, with the assistance of FEMA, have responded to this management challenge by establishing the Incident Command System (ICS). Under this system, the senior responding official assumes the role of Incident Commander (IC) and remains in that position until and unless relieved by a more senior official. There is an established staffing structure to support the IC. The supporting staff structure is based upon functional requirements; the identified functions must be performed regardless of the nature of the problem. The number of people assigned to the various functions will vary depending upon the situation. The IC has three staff functions—safety, liaison and information, and four line functions—operations, planning, logistics and finance/administration. The FBI, due to its different responsibilities, has taken a slightly different approach. The senior special agent becomes the OSC and is in charge, but the organizational structure of the supporting staffing structure is somewhat different. The FBI establishes a Joint Operations Center (JOC) to support the OSC and meet the consequence-management requirements. These plans are further described in Appendix B.

At the federal level, the FBI is the lead agency for crisis management and FEMA is the lead agency for consequence management. As noted in Section 2, the activities are likely to occur simultaneously. At any given time, there can only be one Lead Federal Agency (LFA), defined as "the agency designated by the President to lead and coordinate the overall federal response." The identity of the LFA may change over time. Effective federal command and control depends upon clear designation of the LFA and communication of that designation to all of the responding organizations.

6.5 Communications

Communications between all components of the emergency response will be a vital element in management of the response to the incident. This involves more than having compatible communications equipment, although that is essential. It also includes use of common language and an understanding of what information is essential to the various responders and the public while also not compromising information needed for law enforcement. Fortunately, these are issues that can largely be resolved through the planning and training process. Subject to financial constraints, actions can be taken to assure that equipment is compatible. Training and exercises can develop the use of common language. Essential elements of information can be identified. Procedures and guidelines can be developed to implement a proper balance between protecting that information which must be secured and releasing that information which necessary for an appropriate response. Issues of public communications and information release will be addressed in Section 7.

7. Public Communication

In today's "global village," news about a terrorist event involving radiological materials will instantaneously capture worldwide attention. Responding communicators face a variety of public information challenges. To succeed, they and their fellow crisis- and consequence-management responders must fully embrace an overarching government policy that recognizes public health and safety are paramount. Public information personnel must gather, understand and rapidly release a large volume of incident and response details involving complex technical data. Such information will have a profound impact on the public's reaction to the event and the government's response. This Section discusses public communication issues and provides recommendations for dealing with them during a major radiological terrorist event.

7.1 Communication Policy

A successful consequence-management communications program achieves trust and credibility. It begins with the fundamental objectives of reducing risk to the public and enabling those affected to comprehend the scope of the problem and make informed decisions. Such a program should have a well-defined organizational structure, trained staff, and modern equipment to disseminate timely, accurate, clear and consistent information without creating unwarranted fear. Ultimately the communications program should be viewed as a process of sharing information between response experts and the public. Emergency responders must recognize the short- and long-term information needs of the public, media, and response forces and incorporate that understanding in their plans and actions. The command structure should ensure that these needs are met. However, it must not permit the release of information that might further jeopardize the safety of the public or response personnel.

7.2 Information Management

During any major disaster, one of the challenges is the successful management of the large volume of data flowing into a command center. Compounding the problem are incomplete or inconsistent details that must be analyzed. The designated operations center is primarily responsible for resolving these issues. However, communicators are required to inform the public of the situation and the response. They must gather and sift through available facts to decide what to prepare for release, determine what missing data must also be addressed, and coordinate information for dissemination while other responders are still trying to verify its accuracy and appropriateness. Implementation of a comprehensive public communications plan will greatly assist these efforts.

7.2.1 *Joint Information Center*

An important mechanism to meet these objectives is the early establishment of a Joint Information Center (JIC) when a significant terrorist event appears likely or after an incident occurs without warning. Predesignated, equipped, and trained communicators from crisis- and consequence-management response organizations should be integrated into the JIC, which serves as the focal point for coordination and release of incident-related information to the public and the media. It is also one of the commander's tools to convey information to the response force.

JIC operations and products should truly be a cooperative effort, providing combined local, state and federal information. Depending on the incident location and potential for cross-border effects, response organizations could also include private or nongovernment organizations and foreign government responders. Federal emergency information plans and policies recognize that no single agency at the local, state, federal or private-sector level has the authority or expertise to act unilaterally in response to threats or acts of terrorism involving WMD (FEMA, 1999). Federal consequence-management plans for public information are summarized in Appendix C.

Typically, the JIC director is from the lead organization. Like the ICS, JIC leadership can transition as various response elements arrive on the scene. The U.S. Department of Justice (DOJ), through the FBI, is currently the LFA for crisis management, and it would take charge of the JIC. FEMA is the LFA for consequence management. Ideally, however, during consequence-management operations, the JIC would be co-directed by the lead local, state and federal public information officials to ensure cooperative strategic planning and information sharing. This type of a management structure may dilute responsiveness if the three co-directors cannot consistently form a consensus, plan, and act decisively. In instances where agreement cannot be reached, the state's senior public information officer should take the lead.

The JIC should be located near the operations center, upwind of contaminated areas. As personnel arrive, they must be informed of the situation, the JIC organization, policies and procedures, their responsibilities, key messages, and types of nonreleasable information. Released material should be available for review. Assignment of JIC liaisons to various operational, command and control locations can expedite information exchange and coordination. A suggested checklist of JIC initial actions is provided in Appendix D.

An initial statement on the incident and the response must be provided as soon as possible. Communicators should use a variety of methods and channels to disseminate that statement and subsequent information to the public. However, the media, which plays a key role in any emergency environment, is an efficient and indispensable means of the communicating with the public. The media can inform and educate the public, as well as affect attitudes about the event and the response.

The JIC directors should identify and establish a media briefing area away from site operations, but near the JIC. Responders should brief reporters frequently and regularly as new information becomes available. Key responding organizations should participate in the briefings. Specialist briefings should be considered after, or in addition to, general update briefings. Communicators must also be prepared to facilitate media photo opportunities and interviews that do not unduly interfere with the mission. Opportunities should be sought and offered regularly; their lack of availability may generate distrust of responders among the media and the public.

Commercial television, radio, and the Internet are well suited for the communication of changing information to a widespread audience. In addition, there is the national emergency alert system for citizens immediately affected by the event. Emergency services personnel might also conduct block-to-block notifications if sheltering or evacuation is necessary. FEMA has a radio network and a television channel. Other communication means include:

toll-free call-in numbers for taped radio broadcast announcements, information hotlines, help lines, a specially designated broadcast channel, official photography (still, digital and video) of response activities, live feed capability, newsletters, and meetings.

JIC dissemination of accurate, consistent and timely information that explains potential health risks in easy-to-understand language is crucial. It should, for the short-term: (1) instill confidence that government will conduct response and recovery operations fast, effectively and efficiently; (2) provide critical information about sheltering or evacuation, should this become necessary; (3) indicate how to request assistance; and (4) release authoritative information to deal with unsubstantiated rumors. For the long-term, communicators will want to gain public trust and confidence that contaminated areas, once remediated, are safe.

Successful JIC communication depends upon clearly designated decision makers and release authorities, who should be pre-identified to ensure a rapid flow of all information, but particularly that which impacts health and safety. Plans should focus on conveying protective action and emergency information prior to, during and immediately after a terrorist event. Communicators must provide information on health risks; response and recovery; and impacts on property, the economy, and the environment. These details should increase the public's understanding of the scope of the incident, issues and actions taken by emergency responders to control the site, the status of the response, scientific uncertainties, and other information that contributes to improved public decision making.

While government focuses on scientific and law enforcement data, the public takes into account many other factors. Officials should fully explain their risk management decisions if they want the public to accept them (Covello *et al.*, 1989). Many questions regarding a radiological emergency can be anticipated and for some, suitable answers can be prepared in advance. Samples of these questions are provided in Appendix E.

Plans must also incorporate methods for the JIC to establish a public outreach program to accomplish dialogue with and feedback from the public following a terrorist event. Communication from the public to the involved local, state and federal government organizations succeeds when citizens' concerns are received, reported and understood.

As the situation returns to normal, large numbers of media depart, and the focus shifts to long-term concerns such as medical and hazard monitoring, environmental decontamination, and site cleanup. The JIC staff and leadership could change as personnel involved with short-term response depart with their organizations. Communicators must address the public's need for information on the long-term issues. The public's psychological, social, political and cultural concerns must also continue to be addressed (see Section 5).

7.2.2 *Communication Challenges*

The communications staff may find it difficult to locate and obtain information from busy experts with conflicting priorities. Each responding organization should support public communications by providing information to the JIC. Personnel must understand the importance of including the JIC staff in planning, forwarding data to them, and responding appropriately to their requests for data and briefers. To aid information flow and coordination, the JIC should assign representatives to the operations center. Liaisons may also be established by the JIC in various other response elements; conversely, response organizations may wish to place technical and legal liaisons in the JIC.

Technical experts often find it challenging to convey complex health risk and protection information in easily understandable language and graphics. Further, they spend significant time collecting, processing, and analyzing field data and are uncomfortable releasing details that have not been exhaustively analyzed and confirmed. This can create significant obstacles to JIC objectives of prompt, clear communications. Public information personnel, who may have difficulties with technical information, must work with technical experts to ensure the media understand the data as well as its uncertainty so they can accurately report it to the public.

In the early hours and days of the response, communicators will be challenged to get crucial, technical information out to the public quickly. Release of radiological information must be concurrent with protective action guidance. Lack of information only adds to fear and speculation. Delays in the release of information to allow technical specialists to prepare perfect radiological data and maps of affected areas that are verified by extensive ground surveys are inappropriate. Basic information should be provided on the amount of radiation or radionuclide contamination and areas affected, even if it is qualified as early estimates. The time between information acquisition, packaging and release can be critical.

Good information management is essential in crisis- and consequence-management response. Without it, no response organization can be effective. The JIC is severely affected by a lack of information and inconsistent data. Therefore, communicators must take steps to ensure information is obtained from the best available sources. Knowing which specialist and response element to rely on may be difficult. There will likely be several experts in each field, and organizations with similar or overlapping responsibilities with potentially differing information or opinions. The JIC should ensure the inconsistencies are resolved and information verified.

7.2.3 *Conflicting Concerns*

Command authority cannot permit the release of information that might jeopardize public safety, emergency response personnel or operations. This creates conflicts that are discussed below.

7.2.3.1 *Public Information versus Crime Scene Control.* The media will pursue their First Amendment rights under the Constitution to cover the terrorist event, and media access to the terrorist site will be a key issue. Their desire for coverage can conflict with legitimate security needs and the government's responsibility for protecting public health and safety (Miller, 1982). Although the media will exploit every means to see the site, they will also request and expect government cooperation and assistance. Communication between the JIC and the media is essential to ensure each understands the others' responsibilities and the constraints involved. Initially, law enforcement authorities will limit or deny public and media access to the terrorist site as they restrict access to the crime scene while they conduct their investigation. The media do not have a right to enter criminal investigation or disaster scenes when the public is excluded. However, the JIC staff should work with law enforcement experts to obtain official ground and aerial footage and photos of the site for release to the media. They should also fully explore the possibility of live coverage for certain response venues or activities. The JIC may establish vantage points for media coverage and create media pools (if necessary), to enter the restricted area at an appropriate time.

7.2.3.2 *Public Information versus Health and Safety Concerns.* Site access by the media and public should be initially controlled due to concerns about the possible or actual presence of radiation sources and the potential for the spread of contamination. Although personal safety issues aren't valid reasons to exclude media from the site, they should be informed of any potential exposure risk. The JIC should work with radiation, health, and safety experts to assist media in entering the area safely as soon as possible.

7.2.3.3 *Public Information versus Assistance to Terrorists.* One of the greatest challenges during a terrorist incident involving the dispersal or potential dispersal of radioactive materials is determining how much detail to release to the public without giving the terrorists information that could cause further harm. Public health and safety decisions regarding sheltering and evacuation are interrelated with emergency information plans and actions. Local and state governments have a preeminent role in protecting life, property and the environment in areas not under the control of a federal agency. Public health and safety must be the first considerations in radiological terrorism response, but law enforcement officials should carefully weigh the value of releasing certain information the public may be demanding. They should seek a balance, trying to release enough information to ensure the public is as safe as possible in a situation with unknown variables, while withholding other information as they continue to search for terrorists and radiological devices that could cause additional harm. Decisions about what information is releasable will depend on the situation, consistent with national and operations security. If there are no concerns about multiple radiological releases at the same or other locations, more information may be divulged to the public. Regardless of the type of terrorist event, law enforcement officials must be careful not to release information that would adversely affect any criminal investigation, apprehension and prosecution of suspected terrorists.

7.2.3.4 *Local, State versus Federal Perspective.* Federal authorities are bound by law to withhold certain information (classified, Privacy Act, medical) from release. Differences may arise between what local, state and federal authorities are required to withhold due to the particular laws, policies and practices that affect different organizations. Further, differences may center on interpretations of what best benefits public health and safety, and whether withholding information for national security and law enforcement reasons is appropriate under the specific circumstances. If these differences and the reasons for them are not understood, they may create tensions, delay information release, or fracture the response effort until consensus is reached.

7.2.3.5 *Worst Case Scenarios versus Public Apprehension.* Emergency response personnel should try to convey health and safety information to the public without creating unwarranted fear. They should strive to instill in the public confidence that their actions, carried out by trained personnel, are appropriate and in the public's best interest. When protective action decisions are announced requiring the public to shelter or evacuate due to radiation exposure concerns, citizens will want to know radiation levels in affected areas. Although it is plausible that the release of tentative worst case data will unduly alarm the public, delays in releasing this information are more likely to create public speculation and alarm. The public's perception of the radiation risks, radiation levels, and areas affected could be much worse than response officials' worst case assessments. For example, public speculation might be that an entire state or region could be seriously affected when response officials are only concerned about radiation levels in a few acres. Citizens should receive at least a summary of the information used by officials for sheltering or evacuation decisions. However, it is essential that officials acknowledge their uncertainties. The information should be qualified as initial, projected worst case data to be refined by actual sampling. If the effects are catastrophic, early release of the qualified initial, worst case data becomes even more crucial.

These conflicting concerns illustrate that coordination of public statements during a terrorist incident is vital. Public information personnel must gather and release information quickly, yet they must appropriately coordinate it among response elements prior to dissemination. This ensures that details are accurate, and that intelligence, law enforcement, and operations resources and efforts aren't compromised. All data may not require the same coordination; however, care must be taken so the right organizations have reviewed it. This helps to avoid release of inaccurate or misleading information, details of sensitive operations or intelligence information concerning terrorist activities.

7.2.4 *Information Release Coordination*

The JIC works for the IC, the OSC, or the lead official. The JIC directors should have significant autonomy to release information based upon the lead official's guidance, their knowledge of the situation and relevant laws, regulations and public information policies, practices and procedures. They must, however, take special care to coordinate information to ensure technical accuracy and appropriate security and legal review. Pre-designated release authorities, by position, for local, state and federal information aid rapid information flow. Easy accessibility to these authorities will ensure that critical information intended for the public is not delayed.

Information security review is conducted to prevent the public release of classified or personal information. Information that reveals details that might jeopardize law enforcement operations or endanger lives should be withheld. JIC personnel should incorporate standard security practices in their operations. If the response force practices "security at the source," release of classified information is less likely. Personal data such as home addresses and details about the medical conditions of victims and responders are examples of information prohibited from release under the Privacy Act. Intelligence information about terrorists should also be withheld. In addition, details of criminal investigations into terrorist activities cannot be released. However, information should not be withheld to protect responders from embarrassment or criticism.

If emergency response personnel are contacted by the media for an interview, they should refer the media to the JIC, or contact the JIC themselves. The JIC should advise these personnel how to respond. When an interview is coordinated and approved, the JIC should advise personnel of key messages (Appendix E), nonreleasable information (Section 7.2.4), and media guidelines (Section 7.3.2).

7.3 Ethical Issues

The command authority has an obligation to provide certain information to the public, the media, and the response force. Each group should understand these issues. In addition, the media have their own ethical responsibilities.

7.3.1 *The Public*

Communications at all levels should be proactive and designed to protect public health and safety. There must be a continuous information flow between local, state, federal, private, nongovernment and foreign government response organizations. The public must be told if they face potential radiation exposure. Not only is this a consideration in the pre-incident stage, when there is warning of a potential attack, but it is also important to notify citizens as soon as possible following an unexpected event to prevent any potential additional radiation exposure. Officials may be faced with leaving dead, injured and fatally exposed victims in a large-scale nuclear detonation event so they can focus their efforts on saving those who have not yet received a fatal injury. The public must understand these difficult issues, and the JIC is the vehicle to provide such information to the public directly and through the media.

7.3.2 *Media*

The media must be told if they face a potential radiation exposure risk working at or near the terrorist scene. As they cover the event, the media will pursue their right to gather information, to speak, and to publish. They can be an independent, objective source of information for the public. The media must remain aware of their potential role as a public educators and how they can affect public attitudes and knowledge by the way information is presented (Leone, 1996). Since some terrorists seek publicity, terrorist events present the media with difficult ethical, moral and legal choices. At times the media can become accomplices to terrorists, intentionally or not (Long, 1990). Access by the media to in-progress terrorist events and reporters' portrayals of terrorists can result in strained relations between response elements and the media. They can become part of the incident, not just observers (Simon, 1994). Some media organizations developed guidelines following excessive coverage of terrorism incidents in the 1980s; however, there has been no consensus on the need for guidelines (Task Force, 1986). The media must use their professionalism and common sense to guide their coverage of the event and to keep it balanced and in proper perspective. By following the "do no harm" philosophy espoused by former Ambassador L. Paul Bremer III[16] (Simon, 1994), they can endeavor to cover the unfolding events without harming responders' or victims' lives.

The JIC should explain what type of information jeopardizes operations or endangers lives and request the media's assistance in not reporting that information. Examples of information that should not be reported include:

[16]U.S. Ambassador-at-Large for Counter-Terrorism, 1986 to 1989.

- Extent of intelligence or knowledge about the terrorists. Law enforcement information.
- The effectiveness or ineffectiveness of security measures for the response forces.
- Details of future plans, postponed or cancelled operations.
- Details, photography and imagery that would reveal the specific location of crisis-management response forces.
- Possible consequence-management activities in support of crisis-management response forces, particularly regarding rescue activities.
- Operational or support vulnerabilities that could be used against consequence and crisis-management forces, such as details of personnel losses or equipment damage.

7.3.3 *Emergency Response Personnel*

Emergency response authorities help set the tone and influence how the media and the public will react to a terrorist event (Simon, 1994). They must be aware of the effects of every word they use and every action they take. Visible actions of leadership are important, such as visits to the emergency site.

Media interviews given by emergency response personnel are a valuable means of communication with the public. However, individual members of the response teams have the right to decline to be interviewed by the media. If an interview is given, the individual is responsible for protecting intelligence and other classified information, and sensitive operations details that may have an impact on the safety of response teams and the success of the operation. During interviews, it is appropriate to provide the interviewee's name, expertise and role in the operation. Interviewees should decline to answer questions outside their expertise. Personnel scheduled for an interview should be pre-briefed by JIC personnel to ensure they understand response force information policies and key messages as well as what the media expects of them.

7.3.4 *Commanders*

Commanders must ensure accurate, complete information is released in as timely a manner as possible consistent with legitimate safety and security constraints. When authorization has been provided to allow response force personnel or media to enter areas of increased exposure or other hazardous areas, the commander must ensure that they understand their potential risks. Further, these individuals should be provided with appropriate safety equipment and be informed of guidelines and restrictions.

7.4 Recommendations

Local, state and federal governments should develop coordinated communication plans to inform and educate the entire response (government and nongovernment) community and the public about the nation's vulnerability to terrorist attack before such an event involving radiological material occurs. The awareness campaign should be designed to sensitize the public to potential WMD terrorist attacks, to change public behavior, and to train citizens in appropriate responses before, during and after attacks. Officials must put such a campaign in perspective by explaining the probability of such an event occurring at any given location and the importance of contingency plans.

Local, state and federal emergency response and public information personnel should receive risk communication and media relations training, develop response plans that include public affairs personnel on first response teams, and exercise them. Plans should include as much prepared generic material as possible to save time at the scene; examples include response team missions and capabilities, team key messages, and answers to questions about radiation and contamination. This information can be reviewed, tailored, coordinated, then released, by the JIC. The format should include hard copy, electronic (disk, CD-ROM, Internet), and video, as appropriate.

Senior officials and public information personnel must endeavor to use an appropriate tone and to place the event in proper perspective. Furthermore, they should aggressively pursue communications strategies that enable the public, the media, and the response force to clearly understand the potential immediate and long-term radiological and nonradiological risks and the scientific uncertainty associated with them. This includes training technical specialists to recognize public and media needs and to develop skills to clearly communicate with them.

The lead local, state and federal agency public information officials should plan to establish, staff, equip and operate an JIC near the terrorist scene and the command center. This JIC should serve as the focal point for media, public and response force crisis- and consequence-management information gathering, coordination and dissemination. JIC leadership and staffing should change when the response transitions from crisis and short-term activities to long-term response. The JIC organization must be flexible enough to accommodate the arrival and departure of public information responders without affecting the information flow to the media, public, and response force.

Senior response officials should release contamination levels and exposure information through the JIC to the public when sheltering or evacuation is recommended. Since actual monitoring and sampling information will be initially limited or nonexistent, data from computer estimates should be qualified and used.

Local, state and federal officials should emphasize to their response forces the importance of pushing information to the JIC so it can proactively plan, prepare, coordinate and disseminate information. Public information personnel should be integrated in all emergency preparedness planning to ensure effective communications operations. The JIC must have liaisons in the command center and other response elements to facilitate information flow. Similarly, key response elements may find it expedient to send liaisons to the JIC. Release authorities should be pre-designated and available for information coordination and review so information flow isn't delayed.

Public information personnel should develop policies and plans that encompass the short- and long-term public information needs. These needs must be communicated to all response organizations. Feedback from the public and the response community must be analyzed and acted upon appropriately. Response forces and government organizations should actively embrace the public in decision making whenever possible. This is especially feasible in long-term planning and decision making.

JIC officials should obtain official footage and photographs for release to the media. It should include scenes and operations in restricted and nonrestricted areas. The public information staff should also seek live coverage capability wherever and whenever possible, as well as safe vantage points to accommodate the media at the scene and plan for media pools. Responders should have the technology to video stream media briefings on the Internet.

An Internet template should be developed and preloaded on a federal agency homepage for JIC use in disseminating crisis- and consequence-management information to the media and the public. This capability should be practiced during exercises.

A password protected Intranet template should be developed and preloaded on a federal agency homepage as one means of internally disseminating information to the response force.

The media should establish terrorism coverage guidelines that seek to do no harm. During a terrorist event, they should practice self-regulation. They are entitled to receive an orientation briefing on safety hazards and types of potentially sensitive information so lives are not jeopardized as a result of reporting on the event. The media should be keenly aware of their ability to inform and educate the public, their ability to provide life-saving information to disaster victims, their responsibility to help their audiences evaluate the risks, and to affect public attitude through the way that they present information.

8. Dose Limitations and Guidance

A terrorist act involving the dispersal of radioactive material is by its very nature uncontrolled and possibly unexpected. Under such circumstances it is difficult, if not impossible, to accomplish those actions normally associated with a routine program of practices and constraints. Rather, *interventions* are taken to reduce exposure wherein a set of steps, or *countermeasures* must be taken during such an incident both to restrict doses incurred by members of the general public and to minimize the consequences of unavoidable exposure (ICRP, 1984). The International Atomic Energy Agency (IAEA) has promulgated one such set of internally consistent intervention levels, derived for a generic, international application, which was adopted in 1996 (IAEA 1994; 1996). In the United States, EPA and FDA have derived similar projected dose level recommendations (PAGs) intended to serve as a "trigger" for pathway-specific countermeasures (EPA, 1992; FDA, 1998).

Since it is not possible to provide specific radiation protection guidance for all potential terrorist scenarios, it is important to adopt an underlying philosophy on which decisions should be based. If such a philosophy is recognized and adopted, the coordination of response between local, county, state and federal agencies and departments will be enhanced. The exposure guidelines of this Report are based on the comprehensive and widely accepted radiation protection recommendations of ICRP (1991) and NCRP (1993a).

8.1 Dose Limits for Normal Operations

The goals and philosophy for limiting exposure to ionizing radiation are given in NCRP Report No. 116 (NCRP, 1993a). The following material is taken from that report:

> "The goal of radiation protection is to prevent the occurrence of serious radiation-induced conditions (acute and chronic deterministic effects) in exposed persons and

to reduce stochastic effects in exposed persons to a degree that is acceptable in relation to the benefits to the individual and to society from the activities that generate such exposures.

"The specific objective of radiation protection are:

(1) to prevent the occurrence of clinically significant radiation-induced deterministic effects by adhering to dose limits that are below the apparent threshold levels and
(2) to limit the risk of stochastic effects, cancer and genetic effects, to a reasonable level in relation to societal needs, values, benefits gained and economic factors."

To meet these objectives the NCRP recommended:

(1) the need to justify any activity which involves radiation exposure on the basis that the expected benefits to society exceed the overall societal cost (**justification**)
(2) the need to ensure that the total societal detriment from such justifiable activities or practices is maintained **ALARA**[17] and
(3) the need to apply individual dose limits to ensure that the procedures of justification and ALARA do not result in individuals or groups of individuals exceeding levels of acceptable risk (**limitation**)."

8.2 Dose Limitation and Guidance for Terrorist Events

When the potential exposures from a terrorist event are expected to exceed the limits derived on the basis of normal operations (*i.e.*, 50 mSv y^{-1} for workers and 1 mSv y^{-1} for members of the public), a different approach to limiting the exposures is required.

[17]ALARA is an acronym that conveys the principle that, "In relation to any particular source within a practice, the magnitude of individual doses, the number of people exposed, and the likelihood of incurring [radiation] exposures where these are not certain to be received should all be kept *as low as reasonably achievable*, economic and social factors being taken into account" (ICRP, 1991).

In normal situations, a new source of exposure will be introduced only if there is sufficient benefit to outweigh the attendant risk. For terrorist events, the source of exposure will already exist and the decisions on averting dose must ensure that the actions taken to reduce exposures "do more good than harm." For these situations the ICRP has suggested that "justification is the process of deciding that the disadvantages of each component of interest, *i.e.*, each protective action, are more than offset by the reduction in the dose likely to be achieved" and that, "optimization is the process of deciding on the method, scale and duration of the action so as to obtain the maximum net benefit." In the "intervention" system as applied to terrorist events, the guidance given in ICRP Publication 60 is helpful. In that publication it is noted that the dose limit intended for normal situations "...or any other pre-determined dose limits, as the basis for deciding on intervention, might involve measures that would be out of proportion to the benefit obtained and would then conflict with the principle of justification. The Commission therefore recommends against the application of dose limits for deciding on the need for, or scope of, intervention. Nevertheless, at some level of dose approaching that which would cause serious deterministic effects some kind of intervention will become almost mandatory."

8.3 Exposure Guidance for Emergency Responders

Special individual exposure guidance, often in excess of exposure limits, is required for emergency response operations because the benefits associated with establishing control at the scene of a large radiological disaster are so great. In severe disasters, prompt but well-considered actions can potentially save lives and avert significant harm to the public. However, during emergency operations, the principles of justification and optimization continue to apply.

NCRP Report No. 116 (NCRP, 1993a), *Limitation of Exposure to Ionizing Radiation*, has provided broad guidance for emergency responders:

> "Normally, only actions involving life saving justify acute exposures that are significantly in excess of the annual effective dose limit. The use of volunteers for exposures during emergency actions is desirable. Older workers with low lifetime accumulated effective doses

should be chosen from among the volunteers whenever possible. Exposures during emergency operations that do not involve life saving should, to the extent possible, be controlled to the occupational dose limits. Where this cannot be accomplished, it is recommended that a limit of 0.5 Sv effective dose and an equivalent dose of 5 Sv to the skin be applied, which is consistent with ICRP recommendations (ICRP, 1991).

"When, for life saving or equivalent purposes, the equivalent dose may approach or exceed 0.5 Sv to a large portion of the body in a short time, the workers need to understand not only the potential for acute effects but they should also have an appreciation of the substantial increase in their lifetime risk of cancer. If internally deposited radionuclide exposures are also possible, these should be taken into account."

Workers should also be selected on the basis of their experience in performing required emergency tasks because the time to accomplish a task will likely be reduced, thus helping to minimize worker exposure. The number of workers involved in such tasks should be kept as low as strictly necessary for the tasks to be carried out. Only nonpregnant workers over the age of eighteen should be selected.

Because of the great uncertainty in estimating exposures during the early phase of the incident, and the recognized need to control exposure during potentially sensitive gestational periods, it would be prudent for minors or responders who are pregnant, or are potentially pregnant, to volunteer for service during a later, more controlled phase of the incident. As the availability of medically trained human resources may be at a premium during a mass casualty situation, it is recommended that these sensitive members of the population who may be employed in the health services industry be given responsibilities at medical assistance sites remote from the scene of the disaster.

As with occupational exposures, emergency exposures may consist of external and internal absorbed dose components, as determined by the specific nuclides and exposure pathways involved. These exposures are to be considered once-in-a-lifetime, they should be recorded, but they should not be added to a worker's lifetime occupational dose record. However, exposures incurred during more controlled intervention procedures, usually during the intermediate and late incident phases, should be recorded as part of an occupational dose record.

The OSC should compare the total projected risk to workers from performing a possible action, including both the risks associated with the projected exposure as well as other identifiable risks, with the total risk to be avoided by a population as a result of the proposed action. A decision to act should be based on whether or not an action is justified. It is ethically justifiable to place individuals at risk of harm for the benefit of others if they consent to the risk. There are certain role-related responsibilities wherein an imposition of risk for the ultimate benefit of others may be considered an obligation (*e.g.*, fire fighters, urban search and rescue team members, etc.). It is then necessary that decision makers ensure that no more risk is being imposed on the individual than is necessary. This process includes providing appropriate equipment (*e.g.*, dosimetry, protection clothing), record keeping, and medical follow-up (IOM/NRC, 1999a).

The exposures associated with activities that have been justified should not be treated as a "cut-off" between a dangerous situation and a benign one; keeping exposures below recommended limits, even after application of the ALARA principle, does not guarantee the absence of an increased risk of cancer. Likewise, receiving a dose that exceeds a limit does not guarantee that cancer will develop. Workers who receive an absorbed dose to a large portion of the whole body in a short time should be aware not only of the short-term, deterministic effects that may be imposed by their exposure, but should also have an understanding of their increased risk for the induction of a cancer after some latency period.

8.4 Protection of First Responders

Unless the terrorist attack involving radioactive materials is targeted on a known nuclear facility, it is possible that the radiological aspects of the event may not be recognized by first responders to the scene. Since it is unlikely that all such responding individuals have received the training normally required of workers who are routinely occupationally exposed, it is necessary to establish a mechanism to ensure that these individuals are unlikely to receive an unacceptable level of exposure while at the same time permitting them to perform critical missions during the early phase of a disaster.

For this reason, the NCRP recommends that emergency response personnel or response vehicles likely to be the first to respond to a scene for which there has been some indication that the area may be contaminated with radioactive materials, including the site of **any** explosion, should be equipped with radiation detection equipment that would alert the responders that they are entering a radiologically compromised environment. Further, this equipment should be designed in such a way that it can also alert the responders when an unacceptable ambient dose rate or ambient dose has been reached.

Emergency response personnel assigned to respond to a scene with this equipment should receive training that would include the operational characteristics of the equipment, the operational quantities being measured, and the risks associated with exposures that correspond to the preset levels of the alarms.

The NCRP recommends that an ambient dose rate of approximately 0.1 mSv h^{-1} is a suitable initial alarm level. This is a value significantly higher than natural background so that false positive indications are avoided, but not so high that an emergency responder is likely to receive an exposure that would approach the annual limit for a member of the general public if exposed in areas below this value. It is also an ambient dose rate at which it would be appropriate to establish an initial control point to restrict access for radiological control purposes to any unnecessary persons. The second alarm level, the "turn-around" level, is necessary to permit this initial emergency response team to perform additional time-sensitive, critical missions beyond the point where it is recognized that there is a radiological component to the disaster. The NCRP recommends an ambient dose rate and ambient dose for this purpose would be approximately 0.1 Sv h^{-1} or 0.1 Sv. It is essential, however, that initial responders not proceed beyond the point at which the initial alarm level has been reached unless there is a compelling reason to do so. Such reasons include the rescue of injured persons and time-sensitive actions to regain control of the scene. However, if the first responders include personnel with radiation health expertise and more sophisticated equipment, it is more appropriate that judgments involving higher exposures be made at the scene taking into account all the relevant factors specific to the conditions at the scene.

8.5 Protection of the General Public from Normal Radiation Sources: Exposure Limits

The NCRP (1993a) has recommended that for all sources of ionizing radiation other than medical and natural background, exposure to individual members of the general public be limited to an annual effective dose of 1 mSv. This recommendation is designed to limit the exposure of members of the public to reasonable levels of risk comparable with risks from other common sources.

The NCRP (1993a) has also recommended a maximum annual effective dose limit of 5 mSv to provide for infrequent annual exposures. This additional recommendation was made because annual exposures in excess of the 1 mSv limit, usually to a small group of people, need not be regarded as especially hazardous, provided it does not occur often to the same groups and that the average exposure to individuals in these groups does not exceed an average annual effective dose of about 1 mSv.

These limits were established in order to control the exposure of the public resulting from the legitimate use of radiation sources. That is, the use of radiation sources that are authorized and under the control of responsible individuals. Exposures to members of the public as a result of terrorist activities or radiological accidents are not under responsible control and therefore the concept of exposure limitation cannot be used as a tool to limit public exposure. However, this information will be useful in the final stage when the public will participate in the formulation of public policy in regard to cleanup activities.

8.6 Protection of the General Public from Other than Normal Radiation Sources: Countermeasures

Some form of intervention is almost always required to regain control during or after a radiological emergency. An intervention consists of a set of pathway-specific countermeasures designed to avert as much of a projected exposure to a member of the public as is practicable. Typical exposure pathways as well as the source of the exposure are identified in Table 8.1. Examples of countermeasures include shielding, access controls, sheltering, evacuation, administration of potassium iodide, decontamination, and interdiction of food sources and water supplies (Tables 8.2a and 8.2b). However, intervention to avert doses during the early phase of an incident should be independent of dose projections and actions taken during later phases of the emergency (EPA, 1992).

The goal of any radiation countermeasure is to decrease the total dose to the most exposed members of the public for a given pathway. In addition, and especially during the final restoration

TABLE 8.1—*Possible exposure pathways during a radiological emergency involving a WMD.*

Exposure Pathway	Source
External exposure	Detonation of a WMD Plume Surface contamination and activation products Personal contamination (skin and clothing)
Internal contamination	Plume inhalation Inhalation of resuspended contamination Inhalation or ingestion of personal contamination Ingestion of contaminated foodstuffs Absorption through skin, or injection (as through a wound) of contaminated material

after an incident, reduction of the total collective population dose may be an appropriate basis for enacting countermeasures. It may be necessary to take more than one countermeasure to decrease the projected dose for a particular pathway. The projected doses for each pathway should be assessed separately from all others.

A large-scale radiological incident will require urgent action during the early phase to limit potential casualties. A rapid response will be necessary, as will judgments about justification for action, based on very limited analyses; decision making will be used to form an intuitive response (IAEA, 1989a).

Because almost all countermeasures have an associated risk, the use of a particular countermeasure involves consideration of the risk and benefits with a final decision that should be guided by the application of the principle of doing more good than harm. This decision must take into account all potential risks, not simply those associated with radiation exposure. Because the risk associated with a particular countermeasure depends on the nature of the countermeasure, the population affected and other circumstances unique to the situation, it is not possible to set one generally applicable level of averted dose at which a particular countermeasure is justified. However, it is possible to recommend a range of projected averted doses for which each countermeasure should probably be considered. The upper end of such a range could represent a value of averted dose at which the particular countermeasure is almost always justified. The lower end of the range could be thought of as a value of the averted dose at which the

TABLE 8.2a—*Countermeasures available for each route of exposure listed in Table 8.1 (after ICRP, 1984).*

Exposure Pathway	Available Countermeasures
External radiation exposure from nuclides in the plume	Sheltering, evacuation, control of access
Internal contamination due to nuclides in the plume	Sheltering, ad hoc respiratory protection,[a] administration of stable iodine, evacuation, control of access
External contamination from surface deposited contamination and activation products	Sheltering, evacuation, control of access, decontamination
External radiation from surface deposited contamination and activation products	Sheltering, evacuation, relocation, control of access, decontamination
Internal contamination due to resuspension	Evacuation, relocation, control of access, decontamination
Internal contamination due to personal contamination	Control of access, decontamination
Internal exposure due to ingestion of contaminated water and foodstuffs	Control of food and water and use of stored animal feeds

[a]Ad hoc respiratory protection includes actions such as covering the nose and mouth with a dry or wet handkerchief or washcloth.

countermeasure is not likely to be justified. The ICRP (1993) has recommended such a range of exposure levels for the countermeasures likely to be considered during a major radiological incident. The NCRP agrees with this approach and considers it to be useful for the types of disasters considered by this Report.

Such a range of exposure levels is provided for planning purposes and as a guideline for use in the event of an actual disaster. Final decisions during a disaster must be made after taking into account all relevant, situation-specific information that is available. Because there are risks associated with most countermeasures, the projected averted dose at which countermeasures should be considered is likely to be higher than the recommended exposure limits for the public under normal conditions.

TABLE 8.2b—*Countermeasures (from Table 8.2a) available for each phase of an event (after ICRP 1984).*

Early Phase	Intermediate Phase	Late Phase
Sheltering and ad hoc respiratory protection[a]	Sheltering	—
Administration of stable iodine	Administration of stable iodine	—
Evacuation	Evacuation	—
Control of access	Control of access	Control of access
	Relocation	Relocation
	Decontamination of persons	
	Control of foodstuffs and water and use of stored animal feeds	Control of foodstuffs and water and use of stored animal feeds
	Medical care	
		Decontamination of areas

[a]Ad hoc respiratory protection includes actions such as covering the nose and mouth with a dry or wet handkerchief or washcloth.

8.6.1 *Early Phase Countermeasures*

The early phase is the time from hours to days after initiation of the event, when immediate decisions about responses, based mainly on predictions of meteorological and other conditions and projections of exposures for those conditions are required. A more detailed discussion of the phases of a radiological incident is provided in Section 9.1. Countermeasures may be enacted during the early phase of a terrorist incident based on either a warning of a threatened event or the progression of an actual radiological event. Control of access does not require a specific intervention level because the decision to control the movement of people into and out of an affected or potentially affected area should automatically follow any decision to enact another countermeasure (ICRP, 1984; 1993). Control of access may also be initiated by the lead agency in charge of crisis management for security considerations well ahead of the time when consequence-management decision makers decide to enact a countermeasure.

8.6.1.1 *Sheltering and Respiratory Protection.* Sheltering should be considered an effective countermeasure, with little negative impact on the affected community, if enacted for a short period of time (hours). Longer periods could cause social, medical and other problems such as acute anxiety. Sheltering will generally reduce exposure to external radiation and internal contamination by a factor of up to five (Table 8.3). A reduction by up to a factor of 10 may be afforded by advising people to use ad hoc respiratory protection such as breathing through wet handkerchiefs, towels, frequent showering, etc. Following passage of the plume, internal contamination may be minimized by providing prompt notification so that people might open windows and restart ventilation systems to flush out any radioactive material that may have migrated into the structure. Sheltering will almost always be justified at an averted effective dose level of 50 mSv, with an operational intervention dose range extending below that to 5 mSv, at which time sheltering would almost never be warranted (ICRP, 1993).

TABLE 8.3—*Representative shielding factors for gamma sources in a plume (IAEA 1989a).*

Structure or Location	Shielding Factor[a]
Outside	1.0
Vehicles	1.0
Wooden frame house[b]	0.9
Basement of wooden house	0.6
Masonry home	0.6
Basement of masonry house	0.4
Large office or industrial building	0.2

[a]Shielding factor is the ratio of the effective dose received inside the structure as compared with that which would be received if the structure were not in place.

[b]A wooden frame house with brick or masonry veneer is equivalent to a masonry home.

8.6.1.2 *Administration of Stable Iodine.* Administration of stable iodine (*e.g.*, potassium iodide) can reduce or block the uptake of radioactive iodine in the thyroid of exposed individuals. Such compounds will not reduce uptakes and internal exposures due to other radioactive chemical species. Administration should occur prior to exposure to be most effective and should only occur if an exposure to radioiodines is an actual dose pathway for a given terrorist scenario.

However, because the risk of thyroid cancer decreases with increasing age at exposure (Ron *et al.*, 1995), the efficacy of administering stable iodide to mature adults (the age group from which first responders would be drawn) appears limited. However, administration to pregnant women, neonates, and young children appears to be warranted; the potential risk for the ingestion of stable iodide is smaller than the reduction in risk afforded the concomitant reduction in radioiodine uptake in these sensitive subgroups (Noteboom *et al.*, 1997). The averted thyroid equivalent dose range for operational intervention planning is 0.05 to 0.5 Sv. If the decision is made to administer stable iodine, it should be done as soon as possible, preferably with a few hours after an intake of radioactive iodine (NCRP, 1977).

8.6.1.3 *Evacuation.* Evacuation is the most potentially disruptive of all the early phase countermeasures. If executed correctly, it may afford decision makers the opportunity to avert exposures to the public from all pathways. The ideal time to initiate an evacuation is prior to plume passage; evacuation during plume passage could result in greater exposures than if sheltering were implemented instead. Consideration must be given to special, less-mobile subgroups of the population and authorities should be aware that people may evacuate spontaneously. The ICRP (1993) has estimated that evacuation is almost always indicated if the projected average effective dose is likely to exceed 0.5 Sv within a day, or that an average individual effective dose of 0.5 Sv (or 5 Sv to the skin) may be averted during the evacuation. The operational intervention effective dose range extends from 0.05 to 0.5 Sv over the duration of the evacuation.

8.6.2 *Intermediate Phase Countermeasures*

The intermediate phase is the time from days to months after an event is brought under control, when environmental measurements can be used to assess the need for additional protective actions. Relocation, personal decontamination, interdiction of foodstuffs and water supplies, and medical assistance are all indicated as potential countermeasures during the intermediate phase of an event, in addition to those outlined in Section 8.6.1. The public should be involved in decision making during this phase.

8.6.2.1 *Relocation*. Relocation may be distinguished from evacuation by the time over which the countermeasure is enacted (weeks or months). It may be considered simply as a continuation of an earlier decision to evacuate an area or it may be a countermeasure considered well after the passage of a radioactive plume. The decline in the absorbed dose rate will determine when a relocation countermeasure may be terminated. This decline will be driven by radioactive decay, weathering and deliberate decontamination actions. Social and economic factors may affect the decision to terminate a relocation action as well.

The ICRP (1993) has recommended that an averted average effective dose of 1 Sv will almost always justify relocation (ICRP, 1993). A detailed example of the analysis leading to the conclusion that a prolonged, monthly projected effective dose of 10 mSv would also justify relocation has been provided by ICRP (1993). Relocation may be advisable at either lower or higher effective dose rates.

8.6.2.2 *Personal Decontamination*. Contaminated individuals should be decontaminated. This action should be independent of a decision to evacuate or relocate. Individuals should be instructed to undress, shower and change clothes. The clothes and shoes should be bagged until later when they may be surveyed and decontaminated or discarded. Those with the greatest level of personal contamination should be given priority. Extremely high levels of skin contamination may be treated with special agents under medical supervision (see Section 4).

8.6.2.3 *Interdiction of Food Sources and Water Supplies*. Foodstuffs and water may be contaminated with radioactive materials transferred from several pathways during a radiological terrorist event. Sampling and survey information of water and food will most likely be available prior to the need to make any decision regarding interdiction. Protective actions may be divided into those that directly restrict the consumption of contaminated food and water, and those that limit transfer of radionuclides into the food chain from air, water and soil (ICRP, 1993). Both of these categories lead to actions to control commerce in order to avert exposure. Such countermeasures could be enacted separately for each food category (*e.g.*, Table 8.4).

The FDA has recommended one set of protective action guides for the ingestion pathway (replaces the preventive and emergency PAGs set by the FDA in 1982) (FDA, 1998). These PAGs are for a projected value of 5 mSv for the committed effective dose equivalent, or 50 mSv committed dose equivalent to an individual tissue or organ, whichever is more limiting.[18] These doses should be considered operational intervention levels of dose and have been applied to ingestion due to the total diet. FDA provides a set of derived intervention levels; that is, the concentration of a nuclide present in food that, in the absence of any intervention, could lead to an individual receiving a projected radiation dose equal to the PAG. FDA also indicates that emergency planners may use other assumptions based on actual data to obtain more case-specific derived intervention levels. A set of countermeasures might then be planned for wherein a number of actions would be used to avert much of the projected dose to the most sensitive members of the population.

Simply removing the outer membrane or washing contaminated food is a simple, low risk method of averting exposure when such procedures are practical. Providing guidance on the advisability of reliance on packaged foods is another simple strategy to avert exposure. The interdiction of any single category of food is almost always justified at an effective dose of 10 mSv y^{-1} (ICRP, 1993) due to that category of food. Effective doses well below this value could be justified, particularly for food categories that are not critical. Effective doses well above this value could be justified in the unlikely event that alternative food supplies are unavailable.

8.6.3 *Late-Phase Countermeasures*

The late phase is the time from months to years, when actions may be undertaken to reduce levels of contamination in the environment to allow permanent residence at the site under normal conditions. The late phase may be distinguished by the development and execution of a final plan for the consequences of the disaster. The full plan, as well as the dose-based criteria used to implement it, can only be achieved with the involvement and

[18]The PAGs recommended by FDA are derived for the previous radiation protection quantities of effective dose equivalent and dose equivalent given in ICRP (1977) and NCRP (1987).

TABLE 8.4—*Generic action levels for foodstuffs (adapted from IAEA, 1996).*

Radionuclides	Foods Destined for General Consumption ($kBq\ kg^{-1}$)	Milk, Infant Foods, and Drinking Water ($kBq\ kg^{-1}$)
Cs-134, Cs-137, Ru-103, Ru-106, Sr-89	1	1
I-131		0.1
Sr-90	0.1	
Am-241, Pu-238, Pu-239	0.001	0.001

approval of the public. The plan should include final decisions with respect to disposition of contaminated areas, medical and psycho-social follow-up for injured persons, and any other long-term issues related to the effected sites. Hopefully, the interactions with the public during earlier phases has provided the foundation of trust that is essential for success.

Decisions with regard to disposition of the contaminated areas may encompass some or all of the following objectives:

1. acceptance of an area of restricted access with a plan for continual monitoring and control over an indefinite period of time;
2. establishment of limited or restricted use for activities that may be conducted in an elevated background radiation environment; and
3. full restoration of the site for unrestricted use.

It is conceivable that different areas affected by the disaster could be managed in different ways. For example, it may be desirable to fully restore an urban area by means of a systematic decontamination effort followed by a reconstruction program. On the other hand, a remote site may simply be fenced to control access, permitting natural decay and weathering processes to restore the area over a long period of time.

Regardless of the site restoration objectives, the dose-based criteria as well as the construction activities required to reach them should be consistent with the principles of radiation protection for the public outlined in Section 8.1. That is, the principles of justification and ALARA should be used both to help develop acceptable plans as well as to guide the work required to realize the plans. However, the exposure criteria established through the public consensus process could be higher or lower than the NCRP exposure limits for individual members of the public recommended by the NCRP.

The NCRP (1999) has published screening levels (normalized to an annual effective dose of 0.25 mSv) for contaminated soil that are based on consideration of eight different land-use scenarios including agricultural, suburban and industrial areas. These screening levels may be helpful in planning restoration of a site to the exposure level selected in the final plan.

8.7 Summary

Table 8.5 provides a summary of the NCRP recommendations for dose limitation and guidance. This Table is intended to be used as a tool to help planners and to be useful for decision makers during an actual emergency. Final judgments regarding the selection of any countermeasure must be based in the full context of the circumstances of the disaster.

TABLE 8.5—*Summary recommendations for dose limitation and guidance during a terrorist event involving radiological weapons.*

Classification or Action	Applicability	Limit or Guidance[a]
Full mitigation	General public dose limitation	—[b]
Sheltering	Avert dose to general public	5 – 50 mSv (effective dose)
Evacuation	Avert dose to general public	50 – 500 mSv (effective dose)
Administer stable iodine	Avert dose to children and pregnant women	50 – 500 mSv (equivalent dose)
Any single food category[c]	Avert dose to general public	10 mSv y^{-1} (effective dose)
Relocation	Avert dose to general public	10 mSv $month^{-1}$, 1,000 mSv (effective dose)
Annual limit	Recovery workers (nonemergency work)	50 mSv y^{-1} (effective dose)
Guidance for emergency action	Recovery workers (emergency actions)	500 mSv (effective dose)

[a] When two values are given, the lower value represents the lowest effective dose at which the countermeasure is *likely* to be justified; the larger value represents the effective dose at which the countermeasure is *almost always* justified.

[b]These dose limitation values are obtained through the process of justification and ALARA and the results may be higher or lower than the NCRP limit of 1 mSv y^{-1} for individual members of the public.

[c]FDA has provided guidance for intervention in the ingestion pathway based on the total diet (FDA, 1998).

9. Radiological Consequence Management Considerations

As described in Section 2, the defining moment between crisis management and consequence management begins with the successful, or partially successful, execution of a terrorist act. The primary objectives of the local and state governments during the consequence-management phase of the incident are to establish control of the primary site, limit further damage, protect the public, and ultimately provide for the final disposition of the site.

Many communities have already established plans and pre-positioned resources to deal with a wide range of nonradiological disasters. These plans can generally be augmented to deal with radiological disasters. This Section will therefore focus on the special considerations that must be made when the disaster or terrorist act involves radiation sources.

9.1 Definition of the Early, Intermediate and Late Phases of an Incident

Historically, three time phases have been defined that are common to nuclear incidents (EPA, 1992). Although these phases apply typically to reactor accidents and other radiological events, they can also be applied to terrorist events. These are the early, intermediate and late phases. The transition from one phase to another may be gradual, with some actions from one phase overlapping those in another. The phases are distinguished by the levels of contamination, the location of radioactive contaminants, and the dominant pathways through which effects on people and the environment are manifested (IAEA, 1989b). These characteristics determine the emergency response actions that are typically associated with a given phase.

The early phase starts at the beginning of the nuclear incident. Actions are based on predictions of potential consequences. If possible, precautionary actions should be implemented. This phase continues through any uncontrolled releases of radioactive materials to the environment. For example, in the early phase, the dominant pathways of concern after detonation of an RDD are the inhalation of radioactive material in the plume and/or irradiation from material suspended in the plume. In the case of a nuclear weapon, casualties may occur from the immediate effects of the detonation (blast, thermal radiation, and ionizing radiation) as well as from the contaminated cloud. The early phase following either of these events may last from hours to days.

The intermediate phase begins once the uncontrolled release has been terminated, the contaminated cloud dissipated, and rescue attempts completed. Since major, uncontrolled atmospheric releases have ceased, the major pathways of concern are those arising from freshly deposited radioactive materials. These pathways are direct irradiation from deposited materials, inhalation of resuspended material, and ingestion of externally contaminated foodstuffs, including beverages such as water and milk. Protective actions taken during the initial phase may need to be changed and new actions taken to avert some of the exposure to the public. These may include relocation and food interdiction. The intermediate phase may last from weeks to months.

In the late phase (also referred to as the recovery phase), long-lived radionuclides have been incorporated into the environment and the food chain, and extensive sampling results should be available. Restorative actions are taken to minimize or eliminate the need for previously imposed protective actions, so that normal activities may be resumed in the affected areas. Food and resuspension pathways remain important in the late phase. Full recovery from an incident may take years, depending on the levels of specific radionuclide contaminants. The late phase ends when restrictions are lifted, allowing unrestricted access, residency, and land use (such as farming).

Transitions from one phase to another are generally not distinct. The intermediate phase can be expected to overlap with the early and late phases, both because the shifting of dominant pathways may be gradual and because of time and spatial variations in the deposition of radioactive materials. The phases are a useful tool to conceptualize the changing pathways to man and associated countermeasures after a radiological incident.

9.2 Early Phase

Many of the decisions and actions that are required during the early phase of a radiological disaster are identical to those that would be appropriate for other major disasters. These actions include evacuation of affected areas, establishment of a controlled area around the incident site, and management of injured persons. With that in mind, the following discussion is intended to highlight the specific radiological factors that will have an impact on emergency response for first responders. These actions are expected to be appropriate in most cases for both RDDs and nuclear weapons, except that the standoff distances from the incident site and the number of casualties may be much greater than for incidents involving small amounts of radioactive material.

9.2.1 *Recognition that a Radiological Event has Occurred*

Recognizing that a nuclear weapon has been detonated may be easy simply because of the magnitude of the blast. However, because large conventional explosions can approximate the blast levels of low-yield nuclear weapons, this is not necessarily a reliable indicator. Since the peak temperatures achieved in a nuclear detonation are orders of magnitude higher than for a conventional weapon, reports of flash blindness, skin burns, fires in the line of sight of the explosion but well beyond the immediate vicinity of the center of the blast are all strong indicators that a nuclear weapon has been detonated.

If radioactive material is spread by nonexplosive dispersal mechanisms, it may take a considerable length of time to discover the radiological aspects of the incident. Ultimately, the discovery may occur only as the result of a serendipitous radiation measurement or after radiation-induced symptoms appear in first responders or others. Even if the radioactive material is spread by an explosive, it may still take some time before the radioactive component is recognized. It is therefore important that first responders be provided radiation detection equipment to identify a radioactive component after **any** explosion or when there has been some other indication that the event may have a radiological component.

This detection equipment should include a mechanism that provides two alarm levels. The first is to alert the responders that they are entering a radiological environment that will have to be controlled. As discussed in Section 8, a suitable level for this is an ambient dose equivalent rate of approximately 0.1 mSv h^{-1}. First responders entering such an area should notify their command authority to request additional radiological support. If possible, they should then establish an initial control point to limit further access to the site of the emergency. This control point should be considered temporary pending evaluation by appropriate radiation health personnel. If there is a likelihood of significant casualties, a pressing need to take actions to avoid further threats to public safety or any other compelling reason, the first responders may proceed beyond the initial control point. However, if ambient dose equivalent levels should reach the second alarm level, the first responders should immediately return to the control point to determine a further course of action after consultation with radiation health personnel. As discussed in Section 8, a suitable ambient dose equivalent rate for this second alarm level is approximately 0.1 Sv h^{-1} or when the integrated ambient dose equivalent reaches 0.1 Sv.

The ambient dose equivalent levels for these alarm levels are selected by the NCRP to enable first responders with minimal equipment to perform critical, time sensitive missions without incurring unacceptable risks. If the first responders include personnel with radiation health expertise and more sophisticated equipment, it is more appropriate that judgments be made at the scene taking into account all the relevant factors specific to the conditions at the scene.

9.2.2 *Actions of First Responders*

The initial response force in any terrorist incident would be emergency personnel arriving at the scene. In many cases this would be the fire department, police, or perhaps a first aid team in an ambulance. The initial actions taken by these individuals are critical.

To minimize their risks from exposure to ionizing radiation, all on-scene personnel should carry out their responsibilities keeping in mind three principles to minimize exposure. First, minimize time spent in a radiological environment. Second, maintain the maximum distance from sources of radiation. Third, whenever possible, use shielding to reduce exposure. All personnel responding to the scene of a radiological incident should be given a personal dosimeter and should wear appropriate clothing that will minimize contamination. Medical personnel who will be handling potentially contaminated patients should wear surgical gloves and appropriate anti-contamination clothing. Disposable gowns are particularly useful for medical personnel because they can be easily and quickly changed if necessary as they move from patient to patient.

A preliminary assessment of the anticipated number and severity of casualties must be made as soon as possible so that a sufficient number of medical facilities can prepare to receive casualties. In the event of a large-scale incident involving many casualties, a disaster management approach will be necessary. This includes the use of local hospitals, satellite emergency care clinics that are available in many large cities, the activation of specially trained state disaster teams, the use of military medical assistance (potentially using portable hospitals) and possibly the need to transport casualties to other communities.

The initial responders should collect all ambulatory personnel who were potentially contaminated or exposed in the incident and move them to a suitable, nearby location. If it is practical to do so, these individuals should be separated from each other to avoid further contamination.

Ambulances should be called as soon as significant injuries are identified. Once ambulances arrive, they should be parked upwind from the scene to avoid airborne contamination or exposure to any toxic fumes or smoke from the incident site.

An initial, rapid radiological assessment of the site should be conducted to ensure that conditions are acceptable for the entry of personnel who are qualified to render first aid assistance to nonambulatory patients. As soon as possible, all injured personnel should be evaluated by medical professionals qualified to perform medical triage.

If small areas of very high activity are present, temporary shielding around these sites should be considered. Materials suspected to be radioactive should not be handled directly. This will avoid contamination and reduce exposure (even a small distance between the source and the body can significantly reduce a potential exposure). Collection of evidence and cleanup should only be performed with radiation monitoring support.

9.2.3 *Use of Predictive Models in the Early Phase*

In the case of either a nuclear weapon or a large RDD, significant amounts of radioactivity will be deposited slightly upwind to far downwind of the point of detonation. The time immediately following detonation is characterized by the need to make rapid decisions concerning protection of the public and rescue of victims, often with minimal measurement data to use as a basis for decisions. Reliance is necessarily placed on consequence predictions made prior to or at the time of the event. Models are useful to illustrate the extent of possible consequences. However the lack of knowledge, particularly in the early phase, of important weapon parameters introduces large uncertainties in projected results obtained from any model. For example, the total activity in the RDD or the yield of the nuclear weapon will not be known immediately. Nevertheless, these techniques are essential tools for decision makers during the early phase. One such organization with the mission to provide modeling information is the Atmospheric Release Advisory Capability (contact information in Appendix F). However, detonation of a nuclear weapon or large RDD in an urban area will be much more difficult to model because of the complex geometry. That is, wind patterns in cities can be complex, so fallout may be spotty and difficult to predict. Even in nonurban settings, patterns of radioactivity on the ground will be determined largely by local meteorological conditions and it is therefore impossible to accurately predict deposition levels. Additionally, the locations of people with respect to a contaminated cloud and contamination on the ground will not be immediately known, so individual exposure estimates will have large uncertainties.

9.2.4 *Protective Actions*

Appropriate immediate actions include temporary shelter or evacuation of downwind areas, establishment of a controlled area around the incident site, and management of injured persons. The boundaries of this area may or may not coincide with boundaries established to control the crime scene. Emergency workers should enter relatively uncontaminated areas as soon as possible to help evacuate survivors. However, entry into more highly contaminated areas requires careful judgment and further guidance is provided in Section 8. Areas close to and downwind of the area with established radiological controls should be entered only after personal radiation dosimetry and portable radiation survey meters are available, the need for entry has been established, and the individuals entering the area have been fully informed about the immediate and long-term risks they are taking. Emergency response personnel should check for residual radioactivity as they proceed toward the incident, and not linger in areas with high exposure rates.

Monitoring/decontamination facilities to be used for the general public should be set up at large civic or government facilities such as auditoriums, aircraft hangers, or armories. Although hospitals also could provide decontamination facilities, they may be overwhelmed by casualties. Therefore, general public monitoring/decontamination stations for uninjured persons are best established away from medical facilities. The location of these facilities should be announced using the media or other means of public communications discussed in Section 7.

The early phase protective actions for the public are similar to those specified in many radiological emergency plans used by communities close to a commercial reactor facility. These are primarily temporary sheltering and evacuation and are taken to minimize or avoid exposure to airborne releases of radioactive materials. If some warning is available concerning the location of the device, an evacuation may be possible. For the case with only short or no warning, several actions are possible. Before and after detonation, ambulatory persons should be advised to seek shelter, preferably in the basements of buildings. Passage time for a contaminated cloud will vary depending on the distance from the detonation site and meteorological conditions. Evacuation following the passage of the cloud is desirable if the available shelters do not provide sufficient protection from the exposure. For shelters that provide a high degree of shielding, it may be best to let the fallout decay for several hours or more to reduce exposures during the evacuation.

Sheltering is advisable far downwind during cloud passage. It is unlikely that an orderly evacuation of nearby areas could be accomplished prior to the arrival of a contaminated cloud. In any case, the arrival of the cloud during an evacuation could easily result in more exposure than if persons remained inside shelters. Securing ventilation prior to arrival of the cloud will help minimize contamination of internal systems of downwind buildings. If individuals are in vehicles and caught under the cloud, they should seal the vehicle as tightly as possible (turn off ventilation and roll up the windows) and proceed out of the contaminated area to a decontamination/monitoring facility. If airborne radioactive material is present, the use of shelters for extended periods of time may not be advisable since infiltration would normally reduce their effectiveness.

Because members of the public who have taken shelter generally do not have access to radiation detection instruments, it is essential that regular area-specific guidance be provided regarding further protective actions. As discussed in Section 7, this can be accomplished using the media, the Internet, public address systems, etc. Cellular phones, now in common use, may also be used to help relay evacuation/shelter instructions to the affected public.

Exposure rates from fallout following a nuclear explosion are roughly expected to decrease by approximately $t^{-1.2}$, where t is the time after the explosion, i.e.,

$$I_1 t_1^{-1.2} = I_2 t_2^{-1.2} \qquad (9.1)$$

where I_1 is the exposure rate at t_1, I_2 is the exposure rate at t_2, and t_1 and t_2 are in the same units of time. This is equivalent to the rule-of-thumb that an increase in time after the blast by a factor of seven will result in an exposure rate one-tenth of that measured at the start of the interval. This simple relationship has been derived for situations in which fallout activity is the result of the decay of a complex mix of fission products and is useful for times between 30 min and 200 d after a nuclear explosion (Glasstone and Dolan, 1977). Note that radionuclides dispersed by an RDD will have half-lives much greater than fresh fission products. Therefore, this rule-of-thumb is generally not valid for an RDD. One of the goals of the post-incident monitoring strategy would be to make confirmatory measurements of the rate of decay of deposited radioactive material. For nuclear devices, substantial exposure reductions may be realized by delaying entry into highly contaminated areas.

9.3 Intermediate and Late-Phase Considerations

After the terrorist attack has occurred and the contaminated clouds have dissipated, the major pathways of concern are those arising from freshly deposited radioactive materials. These pathways are direct irradiation from deposited materials, inhalation of resuspended material, and ingestion of externally contaminated foodstuffs. In general, a large number of environmental radiation measurements should be obtained to characterize the areas of concern. Protective actions, such as relocation and food interdiction, may be imposed to avert some of the exposure to the public through these pathways.

Relocation is an intermediate phase action that involves moving people out of contaminated areas and settling them in safe areas on a long-term basis. The decision to relocate a population should be made by local and state authorities based on projected effective dose levels and other factors that may affect the safety of the population. Further discussion of this is provided in Section 8. It is a decision that is made independently of evacuation decisions made in the early phase. That is, persons evacuated in the early phase are not necessarily those that will be relocated on a long-term basis.

Mitigative actions should be evaluated using cost-benefit analyses, where the benefits are primarily long-term risk avoidance. In many cases, it may be advantageous to allow exposure rates to decrease due to weathering and decay, rather than attempt to perform decontamination immediately. Certain high-value structures or facilities may be identified for early restoration. Stakeholders should be involved in restoration planning.

After the contaminated clouds have passed and early phase protective actions have been performed, careful field measurements should be made to characterize the magnitude and extent of radioactive material deposition (fallout). Ground-based measurements will be performed around the periphery, while aerial or other remote techniques may be necessary to safely assess deposition in areas of high exposure rates. Interpretation of aerial surveys may be difficult over complex urban terrain. Reliance may have to be placed on robotic or heavily-shielded, manned vehicles equipped with radiation measurement devices suitable for the radiation environments being assessed. Instruments must be available to measure a wide range of ambient dose rates and contamination levels for the types of radionuclides deposited as a result of the terrorist attack.

During the intermediate and late phases, health physicists will be called upon to perform a variety of tasks. These may include designing, supervising and conducting radiation surveys and sampling programs; ensuring that proper instrumentation and techniques are used for surveys and personnel monitoring; interpreting survey results; projecting exposures to possible residents or consumers; recommending protective actions; identifying important exposure pathways; recommending decontamination techniques (for personnel and facilities); recommending access criteria for entry into controlled areas; and verifying appropriate parameter values are used in pathway models (*e.g.*, resuspension factors). An effective response will require a broad range of health physics expertise and may involve health physicists from numerous organizations.

9.4 Radiological Monitoring and Assessment

The period following the detonation of a nuclear device or large RDD will be characterized by a high degree of uncertainty concerning the spatial extent and magnitude of deposited radioactive material. It is imperative for successful implementation of protective actions that radiation measurements be made as soon as feasible, with due consideration of potential radiological hazards.

While exposure rate measurements are rapidly performed and may be accomplished by personnel with minimal training, assessment of long-term impacts requires knowledge of the radionuclides present and specialized skills. This includes consideration of radioactive decay, weathering processes, land use, and intake pathways. Simultaneous acquisitions of exposure rates and *in situ* high-purity germanium spectrometric measurements (by trained teams either while conducting surveys on the ground or from the air) allow correlation of numerous survey measurements with nuclide-specific determinations of radioactive contamination. If the nuclide mix can be shown to be relatively uniform over an area, exposure rate variations in measurements (corrected to a standard point in time) may be inferred to be due to changes in the total concentration of the mix that is present, assuming equivalent measurement geometries. Simple exposure rate measurements may be used to interpolate between high-purity germanium measurements. *In situ* gamma spectroscopy in areas of high deposition may not be possible due to saturation of the detector. Variable geometry in cities may require the use of collimators on *in situ* detectors, increasing the time needed to survey an urban area compared with an open field.

Unless an RDD includes spent fuel as the source, the released material will probably consist of one or only a few radionuclides. This simple source term is relatively easy to measure and characterize compared with the mixtures potentially released from reactors or nuclear weapons. However, some nuclides that emit principally alpha or low-energy beta particles may be more difficult to detect than mixed fission products. Nuclear weapons will generate a mix of fission products as well as both alpha and low-energy beta emitting radionuclides.

9.4.1 *Radiological Modeling*

Soon after an event, atmospheric dispersion and deposition modeling with an assumed source term may be the only information available to serve as a basis for protective action decisions. As measurements are made, and the confidence in them increases, less reliance will be placed on dispersion/deposition models. Eventually, a reasonably complete description of contamination distributions will be achieved based on actual measurements. Until measurement coverage has been completed, dispersion/deposition models are useful to interpolate between measurement points.

Early in the response, ambient dose contour plots based on models and/or sparse measurements will have a high degree of uncertainty. If these preliminary analyses are released to the media, they should be accompanied with an assessment of the uncertainty and only after carefully weighing the competing needs of timeliness and accuracy. Once released, considerable effort may be needed to retract erroneous information. Section 7 addresses public communications issues in more detail.

Different types of modeling will be necessary after measurements are made. For example, modeling the transport of radionuclides through the food chain is desirable to predict concentrations in foodstuffs. Such modeling, based on measurements of deposition, can provide an initial indication of the need to interdict foodstuffs until direct measurements of levels of radionuclides in contaminated foodstuffs can be obtained. All internal dose assessments require modeling. Derived intervention levels for foods and inhaled activity implicitly incorporate an internal dose model. Modeling of weathering processes that cause spatial redistribution of radionuclides should also be performed. Hydrological models may be used to predict the transport of radionuclides in surface and ground water. Runoff from contaminated areas (or from directing effluents from decontamination efforts into storm sewers that drain to water bodies) may cause contamination or ingestion concerns far from the incident site.

In all cases, the best, most up-to-date models should be used and modeling parameters and assumptions should be confirmed by measurement whenever possible. However, it is recognized that during early stages of an emergency, estimates may be based on simple rules-of-thumb or other rapid assessment tools.

Long-term protective actions (relocation) should be based on projected effective doses with appropriate assumptions concerning time outside/inside buildings, shielding factors for buildings and vehicles, changes in radionuclide concentrations in the environment due to weathering or other factors, land use, intake pathways, or any other significant factors. Confirmation of these assumptions over time is necessary to ensure exposures remain acceptable.

9.4.2 *The Radiological Monitoring and Assessment Center*

The coordination of monitoring efforts will be most efficient if performed from a facility close to the affected area, but out of the fallout pattern. This center would be tasked with:

- the outfitting and deployment of radiological assessment teams
- recording of radiological measurements data
- evaluation of data
- providing recommendations to decision makers
- decontamination and safety monitoring of field team personnel and equipment

In the later stages of consequence management, it may be more convenient to perform data assessment away from the monitoring center that would then function primarily as a data collection center. Remote assessment may also reduce the impact on local resources by reducing the number of people near the incident site. If assessment from remote locations is to be performed, effective communications are needed.

The logistics needed to field a large monitoring and assessment operation are considerable. Additionally, the center will be set up close to the affected area, with potentially thousands of affected individuals needing shelter, food, and possibly medical treatment. Utilities may be damaged or unreliable. Public buildings such as schools and armories may be needed to take care of the affected populace. A convergence of monitoring and assessment personnel should be coordinated and supported so as not to strain local resources. Self-sufficiency in transportation, communications, power and shelter may be needed. Both FEMA and DOD can provide temporary infrastructure and lift capability.

Security should be provided for the monitoring and assessment center. Although assessment personnel need to be able to communicate with the media, they also need to be able to work without interruptions when it is required. Additionally, field team members will need access to the restricted incident site and downwind areas.

Initially, the highest priority would be to obtain many radiation survey measurements to define the extent and magnitude of the hazard from the deposited activity. These measurements may be made with handheld Geiger-Mueller, ionization-chamber, and/or scintillation instruments as appropriate for the nuclides present. A useful list of equipment suitable for radiological monitoring is contained in the Multi-Agency Radiation Survey and Site Investigation Manual (NRC, 2000). The meter reading, type of measurement (open window/closed window), location, and time of the measurement are the most important data to report for each measurement soon after the event. Recording of other data, such as calibration dates, calibration factors, and serial numbers, is less important initially if the meters were in routine service and passed a functional check prior to use. The need to define the level of hazard should instill a sense of urgency: gathering administrative data that may be important in later monitoring should not impede deployment of field teams early in the response.

Upon receipt of radiological data, the monitoring and assessment center personnel should enter it into a database that will facilitate rapid and comprehensive analysis. Early in the incident response when considerable uncertainty exists, the raw data should be sent simultaneously to analysts to avoid processing delays. Later, as the radiological footprint becomes better known and urgency decreases, the processing lag time becomes less critical.

Several products should be developed by the monitoring and assessment center. Initially, map overlays of radiological exposure rate readings will be valuable to those needing to visualize the impacted area. Because fresh fallout from a nuclear explosion decays rapidly, exposure rate readings will need to be decay-corrected to the time(s) of interest. The time until a decay-corrected plot becomes "out of date" will depend on the particular nuclides deposited and the accuracy needed by decision makers. Measurements should be repeated at several locations over time to help establish the decay correction. Otherwise, an established model such as the $t^{-1.2}$ relationship may be used for fresh fission products. Even for radionuclides with long half-lives, concentrations in the environment will change over time due to weathering.

The radiological assessment center should recommend screening values for ground contamination considering land use, such as those in NCRP Report No. 129 (NCRP, 1999). If deposition or concentrations are below screening values, then no restrictions on the distribution of produce need to be imposed. If contamination levels are above screening values, then further assessments of protective actions are necessary.

9.4.3 *Radiological Monitoring Field Teams*

Field teams should be equipped with proper personal protective and radiation monitoring equipment. In most cases, field team members should be issued both a passive dosimeter for record keeping and a direct-reading personal dosimeter to permit each team member to monitor his/her own exposure. Dosimeters provided should have response ranges appropriate for the levels of exposure expected.

Personal protective equipment, such as anti-contamination clothing and respirators, may be needed and should be available. The risks associated with wearing this equipment (*e.g.*, reduced efficiency and heat stress) should be evaluated prior to use. Workers should be monitored for heat stress by industrial hygiene or occupational health personnel.

Field team members should be given guidance concerning their exposure limits, including both radiological and nonradiological turn-back criteria. Monitoring of highly contaminated areas should only be performed if absolutely necessary. Such monitoring may be performed in conjunction with other missions into contaminated areas (*e.g.*, stabilizing a fire or rescuing survivors). Field teams and other responders should minimize the time spent in highly contaminated areas.

Field team members should take precautionary measures to prevent the contamination of equipment or the cross-contamination of samples. Any externally contaminated sample containers should either be decontaminated or double-bagged to prevent the possible contamination of counting systems. Facilities for decontamination of field team personnel, instruments, and vehicles should be provided. Sample activities may range from "very hot" down to environmental levels. It may be prudent to count high and low activity samples in separate counters to limit the possibility of contaminating the low-level counter. Precount screening of samples using handheld instrumentation should be considered.

Caution should be used when employing survey instruments to ensure they are appropriate to measure the type and magnitude of the radiation field. Instruments used normally to measure environmental levels of radioactivity may give erroneous readings in moderately high external radiation fields. In higher radiation areas, it may be necessary to sample and count in a laboratory rather than perform *in situ* measurements. A collimator or shield may also be used to reduce the count rate while improving the spatial resolution when performing *in situ* measurements.

9.4.4 *Data and Sample Archiving*

Surveys should be recorded in an electronic database. Such databases allow manipulation of the data to correct for radioactive decay, and ease the preparation of graphical overlays for display of impacted areas. Staging areas for temporary storage of collected samples should be established.

9.5 Summary

Effective response to a terrorist event involving the dispersal of radioactive material requires recognition of the radioactive component soon after detonation. Because this is difficult without the availability and use of radiation detection instruments, it is recommended that early responders to any explosion be equipped with these instruments and receive appropriate training in their use.

If there are indications of a nuclear yield, responders should use caution when approaching the site of the detonation and downwind areas until radiation measurements have been performed. Similar caution should be exercised approaching the detonation site of a large RDD.

The size of a local radiological monitoring and assessment center must be evaluated on a case-dependent basis. Whether performed locally or remotely, the radiological assessment center must produce products in a timely manner and in forms that are easy to interpret by end users. These data should include assessments of uncertainty, particularly in the early phases of the response. Computer-based information management and summary tools can facilitate processing, transmittal and retrieval of radiological assessment data.

High resolution *in situ* gamma spectroscopy performed in conjunction with exposure rate measurements can be an effective means to efficiently assess radiological conditions over large areas.

10. Planning and Critical Resources

This Section provides basic guidelines for effective planning, including information about how to obtain assistance to prepare to respond to radiological terrorist events. Where appropriate, differences in planning for large metropolitan cities compared with smaller cities or rural areas will be highlighted. Organizations involved with implementation of protective actions during a terrorist event must understand their responsibilities, capabilities and limitations. It is clear that small communities do not have the resources to fully prepare for the types of disasters considered in this Report. Nevertheless, it is important that responsible individuals from these communities consider these issues and establish formal support relationships with larger communities or state governments.

10.1 Considerations for Planning

Effective response requires the performance of several basic functions: emergency command and control, notification and communication systems for responders and the public, emergency assessment, mitigation of hazardous conditions, and protective actions for emergency responders and the public. These functions are described in earlier sections of this Report. The configuration of the emergency response organizations that respond to a radiological terrorist event should be as consistent as possible to the planning that is currently in place for other disasters such as fires, floods, and hazardous material incidents. Understanding the separate responsibilities between local response organizations such as local (*e.g.*, police and fire departments) and federal agencies will be important in preparing an effective emergency response (see Sections 2 and 6).

Determining how much planning is needed and reasonable for a community will be one of the first and most important considerations for emergency planning. Coordination with all responsible organizations is imperative to successfully deal with the consequences of a radiological terrorist event. Dealing with a hazardous material event, including dispersal of radioactive materials, is more complicated than most other emergencies.

10.2 Emergency Response Organizations and Resources

Advance planning is essential; once the incident occurs there will be little or no time for command/control planning. Moreover, if the response is to be effective, every participant must be familiar with the plans. Any differences in understanding will make implementation more difficult and add to the inevitable confusion surrounding the response. All potential participants must train and exercise using the current plans.

Planning is a dynamic process. Plans are constantly being modified based upon changed capabilities, lessons learned during exercises or real events, and emergence of new policies.

10.2.1 *Local Authorities*

In planning for any emergency response, all organizations involved in response, mitigation and recovery must be involved in the planning. These organizations include, but are not necessarily limited to: local government such as the mayor's office, fire department, police department, department of health, department of corrections, hospitals and other medical facilities, department of sanitation, human resources, local Red Cross, transportation companies, and public utilities. Where multiple states could easily be involved in an event, potential jurisdictional problems must be dealt with in advance.

Each local government should designate an emergency planning organization. Large, metropolitan areas will probably have a separate entity devoted specifically to emergency planning and response. Smaller towns and rural communities may not have a separate organization, but should specifically identify the individual or position responsible for a local plan to ensure that initial responders are protected and that mechanisms are in place to acquire assistance in the event of a radiological emergency.

In addition to coordination between local, state and federal organizations, effective emergency planning and response must include coordination with tribal governments that may be close to or co-located with towns and communities.

Recognizing the many demands on an emergency planner as well as the limited resources, smaller towns and communities may find it effective to pool resources to obtain moderate emergency response capabilities. A good place for smaller towns to begin their planning is to contact towns that are located close to a commercial nuclear power plant or a DOE facility. These towns will have tested emergency response plans and perhaps a modest amount of equipment. Understanding what resources these towns have and how they are used can assist an emergency planner in determining how much planning and what type of resources are available to them. The fact that most small communities have volunteer fire departments must also be factored into planning.

10.2.2 *Federal Authorities and Assistance*

The Federal Response Plan (FEMA, 1999) is designed to address the consequences of any emergency situation where there is a need for federal assistance. The plan describes the basic mechanisms and structures by which the federal government will mobilize resources and conduct activities to augment local and state response efforts. Federal assistance will be provided to the affected state under the overall coordination of the Federal Coordinating Officer appointed by the director of FEMA. A list of the federal assets that are available to assist in response to a radiological terrorist event is provided in Appendix F. Section 6, Command and Control, provides background on how the current federal system interacts with the local and state emergency response organizations. Appendix B provides detail on specific planning using the ICS.

FEMA is the primary federal agency designated to assist communities to prepare for and cope with all types of disasters, including terrorist events. FEMA can assist communities in effectively planning for a radiological terrorist event. The FEMA Regional Offices and contacts are listed in Appendix B. Coordination with State Emergency Management Directors is also advisable. A list of the Directors and contact information is also provided in Appendix F.

The Office for State and Local Domestic Preparedness Support, sponsored by DOJ, Office of Justice Programs provides training, technical assistance, and financial support for the purchase of equipment by local and state governments preparing to respond to incidents of terrorism. Contact information is provided in Appendix F.

10.2.3 *Coordination with Tribal Governments*

Since each tribal government can set its own priorities and goals for the welfare of its members, it is important that tribal communities be involved in planning for terrorist events that may occur. FEMA has established an agency policy for establishing government to government relations with Native American tribal governments. DOE facilities have also included tribal governments in emergency management planning for DOE facilities. These interactions provide the basic foundation for further initiatives.

10.3 Emergency Response Planning

10.3.1 *Emergency Plans*

Emergency plans provide the foundation for planning and response to any emergency situation. It is recommended and reasonable to use, if available, emergency plans currently in place. Cities and towns located near NRC licensed commercial power reactors or DOE facilities will already have a good basis for planning and response to radiological and hazardous material emergencies. The requirements for planning for these facilities are found in the joint NRC and FEMA publication: *Criteria for Preparation and Evaluation of Radiological Emergency Response Plans and Preparedness in Support of Nuclear Power Plants* (NRC, 1980). Although this document is specific to nuclear reactors, it provides a good basis for developing and implementing emergency planning for any radiological incident. Many other cities and towns have emergency response plans that can provide the structure for planning for radiological incidents. Appendix G provides examples of two Tables of Contents taken from actual plans from major cities.

When an event occurs that results in the release of radioactive material, the public can best be protected when the response of all organizations is fully integrated. Each organization must have a clear understanding of what the overall level of preparedness must be and what roles each will play. Arrangements between local and state governments as well as any other participating response organizations should be documented in the emergency plan. All formal agreements, such as a Memoranda of Understanding, with other emergency response organizations should be consistent and contain provisions for periodic review to ensure continued applicability. The arrangements should include:

1. Identification of all organizations, their specific responsibilities, and the chain of command that will be followed.
2. Identification of points of contact.
3. Identification of the information that will be provided for initial and follow-up notifications.

The emergency plan is the primary document for providing these details and should describe the duties and responsibilities for each organization responding to a local event. The minimum qualifications for the incumbent of each position should be defined. The lines of authority for key emergency response positions should be established and documented. Emergency plans and procedures should provide for staff augmentation. Procedures should include specific methods and information necessary for timely recall of response personnel. Procedures and checklists should be developed to provide for orderly assumption and transfer of emergency control and coordination functions during the time when the augmentation staff are assuming their responsibilities. At a minimum, each plan should include the following:

1. *Authority, Command and Control*: Describe the primary responsibilities for each response organization, identifying the individual in charge of the response.
2. *Organizational Responsibilities*: Define the responsibility to perform functions during a response, including the development of a concept of operations for each organization that describes its role and how it relates to the other organizations.
3. *Plans and Procedures*: Identify the need for coordinating the development of emergency plans and procedures for each response organization. To ensure effective implementation, plans and procedures must be integrated with all response organizations and routinely evaluated and updated.
4. *Logistic Support, Emergency Supplies, Equipment, Communications and Facilities*: Identify the equipment, supplies, communication and facilities needed to support emergency response tasks.

5. *Training Drills and Exercises*: Establish training requirements for emergency responders and a system of drills and exercises to test response capabilities and demonstrate emergency plan implementation.
6. *Incident Assessment*: Describe means to identify and assess incident conditions and consequences.
7. *Notification and Activation*: Describe means to promptly and effectively inform, activate and coordinate all organizations, groups and agencies which perform emergency response tasks.
8. *Protective Actions*: Identify means to promptly implement urgent protective actions for emergency workers and members of the public.

10.3.2 *Notifications*

Effective planning for radiological emergency response includes planning for all aspects of communications. Prompt and accurate notifications are essential during emergencies to mitigate consequences, activate emergency response organizations and facilities, and notify all organizations responsible for protecting the health and safety of the public. Testing notification protocols and communications on a regular basis is critical to ensuring that all organizations will be appropriately notified and understand the information that is being communicated.

10.3.3 *Equipment*

Effective emergency response requires the ability to access and use reliable emergency equipment. Because radiological detection and protective equipment are highly specialized, often requiring training to ensure effective use, such equipment may not be readily available to many emergency responders. The potential need for this specialized equipment should be assessed based on the potential threat that a local area may have in becoming the target for a radiological terrorist event. If it is determined that the potential threat from a radiological event is extremely low, then emergency planners may decide not to have any specialized monitoring or protective equipment available, but should be ready to immediately request assistance that could quickly and efficiently transport such equipment to their location.

All emergency responders who have the potential for exposure should be issued personnel dosimeters as part of their standard protective equipment. Normal dosimeters should be capable of measuring absorbed doses at emergency levels. It is important to ensure that emergency team members who may enter environments in which they may receive exposures beyond occupational exposure limits have dosimeters or other radiation detection devices that also indicate when a certain exposure level has been reached. A maintenance and calibration program should be established as part of emergency planning program to ensure proper operation of all dosimetric and survey equipment in the event of a real emergency. It may be advantageous, depending on the scope and size of an emergency plan for potential target areas in a community, for emergency planners to elicit the services of a centralized individual monitoring service (IAEA, 1980).

Many factors are necessarily considered in selecting a dosimeter or instrument. Physical factors include radiation type and energy spectrum to be detected, sensitivity, ruggedness, stability and linearity of dose response. The equipment should be insensitive to or protected from environmental factors that include atmospheric temperature, humidity and pressure, dust, vapors and trace chemical contaminants, electromagnetic fields, and mechanical disturbances such as shocks and vibration. Statistical factors include reproducibility and accuracy necessary to make possible assessment of exposures of clinical significance at a later time. Table 10.1 contains information useful in selecting instruments necessary for use in radiological incidents. Neither this Table nor this Section should be considered comprehensive; many resources are available for selection of dosimeters and instruments, as well as for aiding in the design and implementation of a dosimetry plan as part of a larger emergency planning effort.

Most fire fighting protective equipment can be considered acceptable means for respiratory and contamination control in the early minutes of a radiological event; however, without the information obtainable from radiation detectors and dosimeters, emergency responders will not be assured protection against exposure to potentially high levels of radiation. The IC must always be aware of the limitations of the radiological monitoring devices that they may have and understand the information provided by the devices and how the information should be used to protect both the emergency responders and the public.

Selected communication equipment to be used for notifications and reporting includes standard telephones, dedicated phone lines,

TABLE 10.1—*Suggested instruments to be considered during the planning for terrorist events with radiological weapons (after IAEA, 1988b).*

Detector	Radiation Type Measured	Range	Main Use	Advantages	Remarks
Ionization chamber	Beta particles, x rays, gamma rays	30 – 500 mGy h^{-1}	Exposure survey	Almost energy independent	Slow response Low sensitivity
Gas flow proportional counter	Alpha particles	0 – 500,000 cpm	Alpha contamination survey	Efficient	Very delicate Must be used close to electrical ground
Geiger-Mueller counter	Beta particles, x rays, gamma rays	0 – 100,000 cpm	Radiation field measurements Beta and gamma contamination	Rapid response High sensitivity	Highly energy dependent May saturate Radiofrequency sensitive
Scintillation counter	Alpha and beta particles, x rays, gamma rays	0 – 0.2 mSv or 50 – 250,000 cpm	Low-energy contaminants	Rapid response High sensitivity Energy discrimination	Fragile (windowed versions) Highly energy dependent
Portable semiconductor detector	X rays, gamma rays	Spectroscopy	*In situ* gamma spectroscopy	Nuclide identification	Require cooling to liquid nitrogen temperatures – limited time for deployment

Pocket dosimeter including electronic dosimeter	X rays, gamma rays	0.01 mSv and up	Personal monitoring	Inexpensive, portable Some "self-reading"	Provides only integrated dose Some shock sensitive and drift
Thermoluminescent dosimeter, optically stimulated dosimeter, or film badge dosimeter	Neutrons, beta particles, x rays, gamma rays	0.05 mSv and up	Personal monitoring	Can be read only in a processing laboratory	Inexpensive, portable Film provides permanent record
Bubble/superheated drop dosimeter	Neutrons	0.01 mSv and up	Personal monitoring	Inaccurate but reliable	Inexpensive
Electronic dosimeter	X rays, gamma rays	0.01 mSv and up	Personal monitoring	Expensive	Audible signal of field intensity

automatic ringdown circuits (telephones, that when activated will call a list of designated numbers), facsimile machines, paging systems, and perhaps computer data systems. The most important aspect of the equipment is that it be reliable. For this reason, dedicated phone lines, automatic ringdown circuits, and dedicated facsimiles are preferred to regular telephone lines that can quickly be overloaded, especially in an emergency situation. Ringdown circuits are very useful for contacting numerous towns, counties and agencies. All equipment should have operation instructions, qualified operators and identified backup equipment.

The type of communication equipment required should be based on the potential threat for a terrorist event or the need for such equipment for other disaster response. For example, large metropolitan areas, where the threat of a terrorist attack is the greatest, should have the most reliable equipment with adequate backup systems. Areas that may already be prepared for natural disasters may have suitable communication systems with appropriate backup. Smaller, rural areas may not have the resources or need for the more sophisticated equipment. In general, communication equipment should:

1. Be highly reliable primary equipment with backup equipment identified. If practical, equipment should be powered by uninterruptible power sources.
2. Be routinely tested during normal and off-hour periods and be demonstrated during drills and exercises.
3. Be able to handle both voice and data communications as well as teleconferencing capability.
4. Be included in a preventive maintenance program.

11. Training and Qualifications for Personnel Providing Support in a Radiological Disaster

Preparedness or "readiness" to respond to a terrorist event involving radiological materials requires a well-trained response team. Training provides the knowledge by which responders can minimize their own and others' radiation exposure and enable them to make sound decisions to protect health in relation to other, nonradiological hazards. Training must be audience-specific, and focused on developing the skills and expertise to respond effectively to the consequences of a broad array of potential terrorist incidents. It will encompass a range of activities including classroom instruction, "hands-on" training, drills and participation in well-crafted exercises. Due to the many societal misconceptions concerning ionizing radiation, effective training should begin by relating radiation risks to other societal risks or other nonradiation risks faced in emergency response environments, using analogies and language appropriate to the audience. Training must avoid presentations that inappropriately minimize the effects, or conversely, create unwarranted fears. A clear understanding of the risk from radiation in comparison with the risks from competing hazards allows personnel to weigh various risks and make sound decisions that do the greatest good for the greatest number.

11.1 Audiences

Training with respect to the radiological aspects of these incidents must encompass two broad audiences: those having expert knowledge of radiation protection principles (the health physics community), and the general population of emergency responders who have varying levels of familiarity with these principles including:

- firefighters and HAZMAT teams
- emergency medical service personnel (technicians and paramedics)
- law enforcement (including FBI and police)
- emergency room doctors and nurses
- primary/tertiary care doctors and nurses
- mental health, social services, and disaster relief organization workers; psychologists, social workers, psychiatrists, counselors, etc.
- civil affairs personnel
- IC
- senior and public officials of local, state and federal governments

For the audience with radiation protection expertise, training should emphasize the application of that knowledge to the conditions at the scene of a disaster or terrorist incident. For the remaining audiences, conventional emergency response training must be expanded to address the impact of radiation and radioactive contamination in all disaster response activities.

11.2 Depth and Breadth of Training

Training must take into account the many different backgrounds and skills of persons who may respond to large scale radiological incidents. It must be appropriate and relevant to the skills that each audience normally maintains, but adapted to the radiological environments they could encounter during a disaster. The depth of the training should be determined by the trainee's level of responsibility and the role he/she is likely to play. Figure 11.1 depicts the different audiences divided into broad categories of medical, first and follow-on response and command and control populations. The bottom of each pyramid reflects the largest group that requires training, typically performed at an "awareness" level of instruction. Above this are populations requiring more in-depth, "operational" training, focused on skill development. Specialized response assets that have unique, long-term, highly-specialized training requirements, occupy the top of each pyramid.

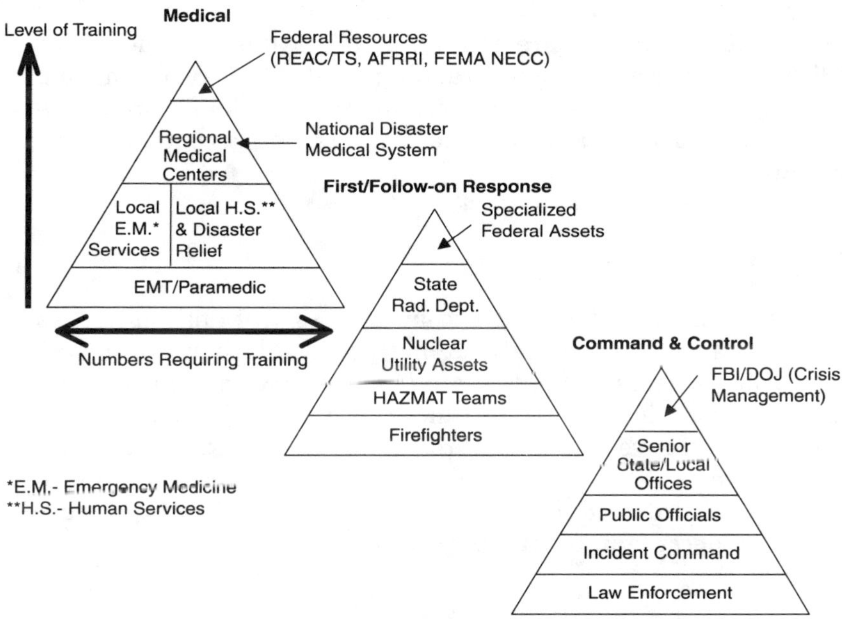

Fig. 11.1. Types of populations requiring training.

To meet these training objectives, broad categories of training have been proposed under the United States Domestic Preparedness Program in an effort to create a large response population with adequate competencies. These training categories are described below (SBCCOM, 2000).

11.2.1 *Medical and Mental Health Management*

This training is highly specialized and specific to emergency and primary health care providers, addressing the medical management of radiation casualties and in managing the contaminated patient. An important subset is the training of health and human services personnel, including mental health professionals, on the unique features and dynamics of radiological incidents.

11.2.2 *First Responder Awareness*

This training is focused on classroom presentation of basic information associated with terrorist response and radiological hazards. Training should be generic, and broadly applicable to all responders. Basic information should include understanding the potential terrorist threat, knowing indicators of nuclear or radiological device use, understanding notification requirements, knowing response plans and procedures, individual protective measures and measures to safeguard others, and decontamination procedures.

11.2.3 *Emergency Responder Operations*

This training builds on basic awareness training and relies more heavily on demonstrations and practice to develop competency at basic skills and operational procedures. Examples of specific operations include using time, distance, and shielding and personal protective equipment to minimize radiation exposure, introduction to radiation detection instruments to measure exposure rates and detect contamination, and the ability to conduct emergency decontamination operations.

11.2.4 *Emergency Technician / Specialist*

This training should include specialized, audience-specific classroom instruction and training using specialized instruments, systems or procedures, augmented by frequent tabletop and field exercises. Training should include incident assessment techniques, advanced implementation of protective measures combined with use of radiation detection instruments and personal protective equipment, and the ability to conduct contamination control and containment operations. Medical personnel require specific training in the emergency and first aid procedures for prompt triage and management of potential radiation casualties.

11.2.5 *Incident Command / Senior Management*

This training is focused on personnel involved with the command and control processes associated with crisis and consequence management. This training should emphasize the complex interactions of the numerous agencies and assets available to respond to terrorism, implementing necessary response plans, public communication skills, and understanding basic technical issues relevant to crisis and consequence management. This audience includes the public affairs official or spokesperson for senior management with responsibilities for communicating with the public. Training for these individuals will include effective public communication and other principles identified in Section 7.

11.3 Training Content

Training of the first responder population (including firefighters, police, and emergency medical services) for radiological incidents should emphasize critical concepts for self-preservation and effective casualty management. Training should include both classroom instruction on basic concepts and principles, hands-on demonstrations of required skills, and drills to reinforce basic procedures. Typical subjects relevant to radiological incident response should include the following (IAEA, 1989a, IOM/NRC, 1999a).

Training for first responders:

- An understanding of the threats and types of radiation exposure:
  - potential terrorist scenarios involving radiation or radioactive material
  - types of radiation
  - differences between radiation exposure and radioactive contamination
  - internal and external exposure
- The health risks posed by exposure to ionizing radiation:
  - deterministic and stochastic effects
  - genetic effects and effects on the unborn fetus
- General radiation protection principles:
  - justification, ALARA and limitation
  - external exposure: time, distance, and shielding
  - internal exposure: respiratory protection, hygiene, and monitoring
- Relevant emergency procedures, including reference or action levels
- Use of instruments and equipment to:
  - identify sources of radiation emission and radioactive contamination
  - measure radiation fields or assess contamination levels
  - monitor individual exposures
- Contamination control and decontamination procedures
- Management of radiation casualties, including mass casualty triage
- Command, control, communication and coordination

Often, the radiological aspects of the management of the type of disaster discussed in this Report may be appended to existing training programs for emergency responders to nonradiological disasters.

Specific individuals or teams of individuals from a given organization will typically be responsible for developing special skills or knowledge to enhance the organization's effectiveness in terrorist response. For health physicists, advanced knowledge or skills that have relevance to effective terrorism consequence management include the following.

Training for health physicists:

- medical effects of ionizing radiation
- atmospheric and environmental transport of radioactivity
- environmental sampling techniques
- assessing exposure and risk from radioactivity in the environment
- ALARA training
- respiratory fit testing and personal protective equipment selection
- contamination control and decontamination procedures
- radiation survey instrumentation calibration, use and maintenance
- radiation detection, measurement and interpretation
- external dosimetry and exposure assessment
- risk communication and media training
- radiation shielding
- internal dosimetry and *in vitro/in vivo* measurements

The training needs of medical care providers, public officials, and senior command personnel are much more specific. Section 4 provides an outline of current medical treatment strategies for a wide range of radiological injuries. Appendix H provides a detailed table of federally supported courses relevant to nuclear or radiological terrorism response.

11.4 Training Frequency and Refresher Training

Training should assure that expertise and proficiency are maintained at a consistently high level. This requires frequent training, exercise, and productive feedback to maintain skills. The need for refresher training depends on how the training is reinforced in daily work practices. Less frequent use of material requires more frequent refresher training. Since the requirements of radiological disaster response will be highly demanding, stressful and complex, and the skills required are not routinely utilized by most responders, refresher training should be performed at least annually to maintain proficiency.

Individual advanced training in a given subject area is generally only acquired once during a career through attendance of continuing education short courses, workshops, conferences, or college courses. Competence within an organization is commonly maintained by rotating personnel through these courses, such that one or more individuals have received training within 1 to 2 y.

11.5 Exercise Requirements

Skills and knowledge critical to effective crisis and consequence management cannot necessarily be maintained through routine job performance or on-the-job training. The best mechanism to assure proficiencies are maintained is through execution of frequent, well-organized exercises involving a number of potential disaster scenarios. Exercises themselves will vary in extent, focus and type depending on the specific proficiencies to be assessed and developed, as well as on what resources are available. Issues requiring attention in planning exercises include the following:

- personnel who will be trained (firefighters, command and control personnel, health care providers)
- definition of training objectives
- available resources (time, training areas, equipment, money)
- the type of training exercise most appropriate to accomplish each objective, given the constraints of available resources
- performance critique and lessons learned

11.5.1 *Types of Exercises*

Exercises are generally described as being of one the four following types.

11.5.1.1 *Map or Tabletop Exercise.* This exercise involves a large room with maps, overlays, flip charts, or other audiovisual material to simulate a scenario and allow for discussion between players. A map or tabletop exercise (MAPEX) is particularly useful in allowing commanders and senior officials to reflect on the overall response to a crisis, including resource identification, logistics requirements, and command, control coordination, and communication issues. A MAPEX allows training in:

- team-building
- identification of resource requirements
- communications/information exchange
- decision making
- planning
- coordinating diverse response activities
- command and organizational relationships.

A MAPEX can also be useful in the planning process for more complex field exercises.

11.5.1.2 *Command Post Exercise.* This form of exercise uses a subset of responders that may be involved in a MAPEX, and can be carried out either in a tabletop fashion, or in an actual field environment. The purpose is to allow the IC, public officials, and senior decision makers to evaluate the adequacy of plans and resources; assess command, control, communication and coordination; and train officials, staffs, and subordinate commanders in crisis- and consequence-management leadership.

11.5.1.3 *Situational Training Exercise.* This type of mission- or scenario-related exercise is limited and designed to train personnel in one collective task, or a group of related tasks, through practice. The situational training exercise (STX) teaches the standard, preferred method for carrying out the task. Examples include patient assessment, extraction, transport and decontamination, or methods for surveying and sampling contaminated environments. A STX can be used to evaluate lessons learned in MAPEX and command post exercises (CPX) to assure high proficiency in command and control and to prepare individual groups or audiences for participation in larger scale exercises.

11.5.1.4 *Field Training Exercise.* These exercises are conducted under simulated incident conditions in realistic scenario environments. The field training exercise (FTX) fully utilizes many or all response elements that would actually participate in crisis-

or consequence-management operations under real-world conditions. Due to resource, logistics and planning considerations, FTXs are less frequent than other exercises, but are invaluable in allowing all responders to coordinate activities, and test command, control and communication systems. The exercises also build teamwork and cohesion and allow the assessment of the logistics infrastructure necessary to support large disaster-response activities.

11.5.2 *Environments and Scenarios*

Historically, domestic nuclear or radiological incident training has focused on two basic scenarios: reactor accident exercises and DOD/DOE sponsored nuclear weapons accident exercises. These scenarios have typically involved a limited number of responders in less populated areas. Terrorist acts, in contrast, will more likely involve highly populated areas in cities and entail a massive response effort. Field training exercises should thus include a focus on urban environments.

The frequency at which exercises are conducted should be such that the interval is both appropriate and consistent with economic factors. Intervals for field training exercises should be no longer than 2 y, with each emphasizing new or different scenarios. MAPEXs, CPXs and STXs should be conducted more frequently as time and resources allow.

11.5.3 *After Action Review, Evaluation and Feedback*

All exercises should conclude with a review of the activities that occurred and a summary of lessons learned. This evaluation should reinforce positive aspects of the response and reveal shortcomings in plans, personnel, training, equipment and procedures so that they can be corrected. Figure 11.2 illustrates the complete training/exercise cycle.

11.6 Trainer/Student Certification, Qualifications and Records

Trainers should have formal education in the subject material above the level of their students, and have sufficient experience with the subject that they can answer questions on theory, practical application, and rationale for the subject. In most cases this will require expertise in both radiation protection and emergency

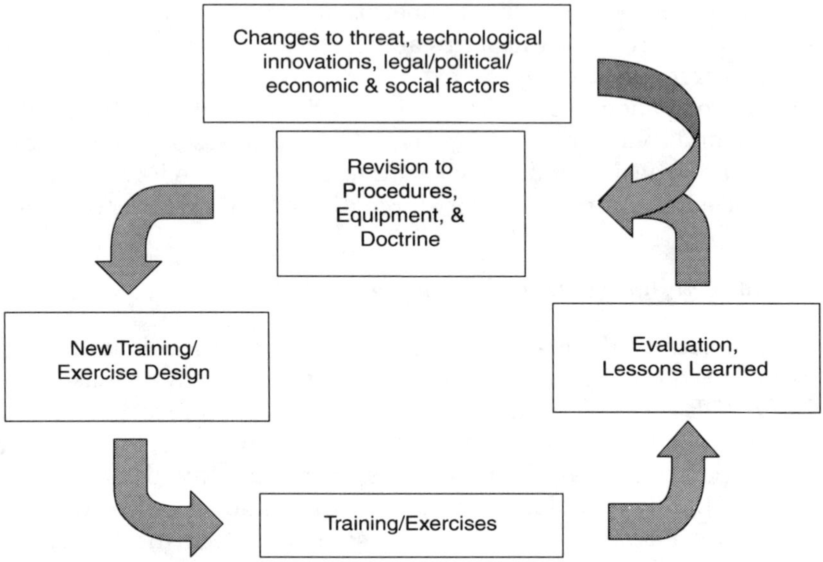

Fig. 11.2. The training/exercise cycle.

response. However, trainers with a specialized knowledge in a specific area may provide training in the context of a curriculum under the supervision of a director or colleague with a broader scope of expertise.

Basic training should be required for those likely to be involved with a role in crisis or consequence management following a terrorist incident involving radiological materials. Persons who have not received this training should not enter environments with significant radioactive contamination or radiation fields except when absolutely necessary and only under the immediate supervision of a qualified individual. Recommended content for this training is provided in Section 11.3. This training should be documented, and renewed periodically as discussed in Section 11.4. Examples of basic training (by specific audience) developed by the Domestic Preparedness Program under the National Defense Authorization Act of 1997 (also known as the "Nunn-Lugar-Domenici legislation") are provided in Appendix H.

A system for maintenance of training information should be used to evaluate the adequacy of the program, establish an inventory of responder skills, and delineate additional training requirements for workers. Records of trainee performance should be maintained by the organization that provided the training and made available to the student. Lesson plans, worksheets, sample exams, and problem sets should be maintained for periodic review.

11.7 Recommendations

Appropriate authorities should establish a training program, including exercises, for those called upon to respond to incidents involving radiological terrorism. This authority should specify what "basic" training is required prior to individuals being "qualified" to respond, and the frequency at which this training must be renewed.

To ensure that standard procedures and methods are used in terrorist crisis and consequence management, appropriate authorities should develop and distribute guidelines on the preferred methods to perform specific missions and tasks, including the use of standardized equipment.

Training and exercises should include environments that are likely terrorist targets, particularly high population areas. Appropriate authorities should assure that "lessons learned" from large exercises are distributed to all relevant agencies, incorporated in response plans and future training, and utilized for the development of other exercises.

There is clear lack of appropriate training for dealing with the social and psychological consequences of incidents involving WMD. Such courses should be developed to augment conventional disaster mental health training, and made available nationally.

Training of key crisis-management personnel, *e.g.*, senior local FBI officials and likely special agents in charge, should include awareness of consequence-management issues and operations.

12. Research and Development Needs

To address radiological terrorist incidents more effectively, a number of areas have been identified in which additional research and/or development is indicated. The major research and development areas are summarized below. It should be noted that many sections include a more extensive summary and that not all of the research and development recommendations are included here.

12.1 Radioprotection and Biodosimetry

Pharmacological strategies for use in large-scale radiological emergencies exist and should be considered (as indicated in Section 4.4) for use in response to large-scale radiological terrorist incidents. The possibility exists that radiation injury could be reduced or repopulation of key cellular systems could be enhanced by administering certain pharmacological or nutritional agents before and after irradiation. However, there is a need for continuing research in this area to understand better the long-term risks/benefits of such a strategy in large, exposed populations. In addition, the effectiveness of radioprotectant drugs administered either before or after the exposure must be better understood.

The FDA has not approved any drugs for prophylaxis against the effects of ionizing radiation. Research data must be acquired to provide a basis for soliciting FDA approval of such an approach. Part of the national preparedness must include obtaining these approvals.

In addressing the kinds of terrorist incidents considered in this Report, it is clear that many of the individuals exposed will not be wearing dosimeters. This prospect could include many of the first-responders. Biodosimetry research is progressing rapidly and the lower limit of detection of some of these techniques is being reduced. Additional and continuing research is necessary to better understand the power of these techniques and to be able to use them more effectively in evaluating retrospectively the exposures of large populations.

The need for long-term medical follow-up of the exposed population has been identified as an important part of the overall response to a radiological terrorist incident. This follow-up would include first-responders as well as the general population exposed to ionizing radiation. At present, there are only a few screening tests for cancer available that are accurate and effective. Additional research is necessary to identify and establish appropriate cancer screening tests. The availability of screening tests could play an important role in the medical management of long-term effects of a terrorist incident.

12.2 Instrumentation and Dosimetry

The availability of appropriate and affordable radiation detection equipment is key to being able to identify and evaluate terrorist incidents involving radiation and radioactive material. These instruments must be rugged and reliable, must be easy to use and interpret, and readily available. It is likely that the types and ranges of the instruments required will change as the transition is made from crisis management to consequence management to long-term remediation. Instrument needs include not only the appropriate survey instruments but also individual dosimeters such as pocket ionization chambers and/or electronic dosimeters.

Additional research and development is required in the area of instrumentation and dosimetry. For example, it is very likely that first-responders will have very little indication that a radiological terrorist incident has occurred. A simple instrument should be available to be permanently mounted in all vehicles likely to respond to an explosion in order to warn of the existence of a radiation field. This instrument could provide a "warning" and a "go–no go" indication to the responders. Such an instrument should be very rugged and reliable. It should require little maintenance, be easy to calibrate and include a simple operational check. Currently, such instruments are not available but they could be extremely important in a real emergency.

Other types of useful instruments are not widely available. For example, there are few high dose and dose-rate instruments available that are suitable for use in responding to terrorist incidents. Normal survey and contamination monitoring instruments are likely to saturate quickly and provide either misleading or no information to the responders. The development of high dose and dose rate instruments that are simple, rugged and reliable is needed.

The exposures received by first-responders will be important for a number of reasons, including planning the appropriate use of key personnel in an extended emergency situation. Thus, reliable personal dosimeters are required. These dosimeters could be simple pocket ionization chambers, although the use of these devices has decreased significantly over the last several years. Electronic dosimeters could be used but there is a need to understand better the response characteristics of the devices, their display ranges, their failure modes, etc. As with other radiation detectors, these dosimeters must be simple, reliable, and widely available at a reasonable cost.

Along with the development of instruments and dosimeters, there is a need to develop expedient survey methods to be used to validate crisis- and consequence-management approaches as well as prediction models, etc. These survey methods also should include remediation activities. A significant body of knowledge is available in the decontamination and decommissioning literature. This information should be reviewed extensively to ascertain its applicability in these emergencies. Documents should be prepared, based on these data, for use in radiological terrorist incidents.

12.3 Psychosocial Aspects

As stated in Section 5, few emergency plans at the local, state or federal level include appropriate consideration of the psychosocial aspects of a radiological terrorism incident. Even though there is some information on psychosocial interventions used after radiological incidences, there has not been a systematic effort to provide an analysis of these situations. It is important that such interventions be recorded, studied, analyzed and evaluated to provide valuable insights relevant to service delivery after a radiological terrorist incident. These considerations must include the early, intermediate and late phases of the incident.

Additional research is needed on the reaction of the general population to decontamination situations, the psychological effects of undergoing decontamination, and approaches to reduce the impact of these situations. Additional attention of the research community must be directed toward the problem of stigma post-incident. In particular, there is a need to develop a better understanding of approaches to prevent or ameliorate the stigma associated with radiation exposure and/or contamination situations.

Research on high-risk groups should be given priority. Some important questions to be addressed include the following. What measures should be taken to reduce or prevent psychosocial effects in emergency workers, evacuees, cleanup workers, and other groups already identified as being at increased risk for such effects? How are children affected by a WMD incident and what special measures and/or interventions can help protect them in such situations? What interventions are most effective in mitigating and/or preventing psychosocial effects in pregnant women and mothers with young children?

13. Summary of Recommendations

Listed below is a summary of the major recommendations made regarding radiological terrorist incidents. Many more specific recommendations and more detail may be given in the text of each specific section of this Report.

13.1 Recognition Capability

Unless the terrorist attack involving radioactive materials is targeted on a known nuclear facility, it is possible that the radiological aspects of the attack may not be recognized by first responders to the scene. Since it is unlikely that all such responding individuals have received the training normally required of workers who are routinely occupationally exposed, it is necessary to establish a mechanism to ensure that these individuals are unlikely to receive an unacceptable level of exposure while at the same time permitting them to perform critical missions during the early phase of a disaster. For this reason, the NCRP recommends that emergency response personnel or response vehicles likely to be the first to respond to a scene for which there has been some indication that the area may be contaminated with radioactive materials, including the site of **any** explosion, should be equipped with radiation detection equipment that would alert the responders that they are entering a radiological environment. Further, this equipment should be designed in such a way that it can also alert the responders when an unacceptable ambient dose rate or ambient dose has been reached. Suggested alarm levels are discussed in Section 8. These systems need not be complex but they should be rugged and reliable.

13.2 Command and Control

In responding to these incidents it is anticipated that effective response may be hampered by a lack of coordination and control, especially early in the response. It is essential that all responding organizations, even if they have independent authority to act, must coordinate their actions with other responding organizations. During terrorist events, when a national level response is warranted, federal authorities are clearly responsible to take the lead during the crisis-management phase of the incident. However, it must also be clear that, during the consequence-management phase of the incident, the responsibility for public health and safety rests with local or state authorities. Federal authorities are expected to play only a supportive role.

13.3 Communications

It is imperative that clear communications be established with the public regarding these incidents. These aspects are discussed in detail in Section 7. The public should be fully informed of the projected impact of the incident as soon as possible after the incident. It will also be important to include in these projections clear statements of the uncertainties associated with these projections. Information should be withheld only if the consequences of releasing the information would adversely impact the ability to protect public health and safety or if the information would aid the terrorists during the crisis-management phase of the incident.

13.4 Psychosocial Aspects

In consequence-management planning and execution, greater consideration must be given to social and psychological issues. This should not only include attention to immediate psychosocial impacts; it should also involve efforts to prevent and ameliorate the wide range of longer-term psychosocial effects that could be expected after a radiological terrorist incident. Section 5 of this Report discusses psychosocial aspects in more detail.

13.5 Medical Response

As discussed in Section 4, medical response in these emergency situations may be hampered not only by a large number of casualties but by an inordinate fear of radiation, radioactive materials, and contamination. It must be made clear in all phases of training that contamination is never immediately life-threatening and that other considerations take precedence over decontaminating survivors.

Although the prompt administration of potassium iodide (KI) is effective in reducing thyroid exposure to radioiodines, based on recent meta-analysis of epidemiological studies, the value of administration of KI to adults is small. As a result there is little reason to consider large programs to distribute KI to adults in the event of terrorist incidents. However, experience from a number of epidemiological studies as well as more recent experience from Chernobyl indicates that the thyroid of the fetus and child is likely to be quite sensitive to induction of thyroid cancer following radiation exposure. As a result, there is reason to have a plan to distribute KI to pregnant women and to children if a terrorist scenario is suspected to involve a nuclear weapon or some other major release of radioiodine.

13.6 Exposure Guidance

The exposure to emergency responders should be limited, if possible, to occupational exposure limits of NCRP Report No. 116 (NCRP, 1993a) (summarized in Section 8). However, during a severe disaster, because prompt but well-considered actions can potentially save lives and avert significant harm to the public, exposures beyond these levels may be authorized. Even in these cases, principles of justification and ALARA always apply for emergency responders.

In order to protect the public, some form of intervention is almost always required to regain control during or after a radiological emergency. Because all countermeasures have an associated risk, the use of a particular countermeasure involves a risk/benefit analysis, and any decision should be guided by the application of the principle of doing more good than harm (Section 8). This decision must take into account all potential risks, not simply those associated with radiation exposure. Because the risk associated with a particular countermeasure depends on the nature of the countermeasure, the population effected and other circumstances unique to the situation, it is not possible to set one generally applicable level of averted dose above which a particular countermeasure is justified. However, it is possible to recommend a range of projected averted doses for which each countermeasure should probably be considered (Section 8.6). The upper end of such a range could represent a value of averted dose at which the particular countermeasure is almost always justified. The lower end of the range could be thought of as a value of the averted dose at which the countermeasure is not likely to be justified. Such a range is intended only to provide guidelines for planning purposes and in the event of an actual disaster. Final decisions during a disaster must be made after taking into account all relevant, situation-specific information that is available. Because there are risks associated with most countermeasures, the projected averted dose at which they should be considered is almost always higher than the recommended exposure limits for the public.

13.7 Late-Phase Decision Making

The late-phase response will include cleaning up the area and restoring it to a preexisting condition. The area to be restored may be quite large as will the cost and effort required to accomplish these tasks. Criteria to be used for cleanup and release of the area for "free use" must be based on agreed upon levels. In selecting these cleanup levels, it is essential that the public be fully involved and that they be full participants in these decisions. Factors such as total cost, time to accomplish the tasks, risks associated with cleanup criteria, etc. will be important parameters in such a decision-making process.

Appendix A

Medical Aspects of Radiation Injury (0 to 30 Gy)[a]

Dose	Initial Symptoms	Initial Symptoms Interval Onset - End	Medical Problems	Indicated Medical Treatment	Disposition without Medical Care	Disposition with Medical Care	Clinical Remarks
0 – 0.35 Gy	None	N/A	Anxiety	Reassurance. Counseling at redeployment	Home/Duty	Home/Duty	Potential for anxiety
0.35 – 0.75 Gy	Nausea, mild headache	ONSET: 6 h END: 12 h	Anxiety	Reassurance. Counseling at redeployment	Home/Duty	Home/Duty	Mild lymphocyte depression within 24 h
0.75 – 1.25 Gy	Transient mild nausea. Vomiting in 5 – 30% of personnel	ONSET: 3 – 5 h END: 24 h	Potential for delayed traumatic and surgical wound healing. Minimal clinical effect	Debridement and primary closure of any and all wounds. No delayed surgery	No further radiation exposure or elective surgery	No further radiation exposure	Moderate drop in lymphocyte, platelet and granulocyte counts. Increased susceptibility to opportunistic pathogens

1.25 – 3.0 Gy	Transient mild to moderate nausea and vomiting in 20 – 70% of personnel. Mild to moderate fatigability and weakness in 25 – 60% of personnel	ONSET: 2 – 3 h END: 2 d	Significant medical care may be required at 3 – 5 weeks for 10 – 50% of personnel. Anticipated problems should include infection, bleeding, and fever. Wounding or burns will geometrically increase morbidity and mortality	Fluid and electrolytes for GI losses. Consider cytokines for immunocompromised patients (follow granulocyte counts)	$LD_5 – LD_{50}$ No further radiation exposure, elective surgery, or wounding. May require delayed evacuation from the scene	No further radiation exposure, elective surgery, or wounding	If the lymphocyte count is greater than 1.7×10^3 μL^{-1} 48 h after exposure, it is unlikely that an individual has received a fatal dose. Patients with low (300 – 500 μL^{-1}) or decreasing lymphocyte counts or low granulocyte counts should be considered for cytokine therapy. A biological dosimetry assessment using metaphase analysis or other techniques should be done.

Dose	Initial Symptoms	Initial Symptoms Interval Onset - End	Medical Problems	Indicated Medical Treatment	Disposition without Medical Care	Disposition with Medical Care	Clinical Remarks
3.0 – 5.0 Gy	Transient moderate nausea and vomiting in 50 – 90% of personnel. Early: Mild to moderate fatigability and weakness in 80 – 100% of personnel.	Nausea / vomiting ONSET: 2 h END: 3 – 4 d Diarrhea ONSET: at 10 d END: 2 – 3 weeks	Frequent diarrheal stools, anorexia, increased fluid loss, ulceration, death of crypt cells and Peyer's Patch lymphoid tissue Increased infection susceptibility during immuno-compromised time frame Bleeding diathesis at 3 – 4 weeks due to megakaryocyte loss	Fluid and electrolytes for GI losses Consider cytokines for immunocom-promised patients (follow granulocyte counts). Specific antibiotic therapy for infections May require GI decontamination with quinolones. Use alimentary nutrition	$LD_{50} – LD_{100}$ Survivors may be able to return to light duty after 5 weeks. No further radiation exposure. May require delayed evacuation from scene	Increased percentage of survivors may be able to return to duty after 5 weeks. No further radiation exposure. May require evacuation from scene for adequate therapy.	Moderate to severe loss of lymphocytes. Follow counts in first few days if possible for prognosis. Moderate loss of granulocytes and platelets. Hair loss after 14 d, Thrombocytopenic purpura appears after 3 weeks. Consider cytokine therapy and biological dosimetry assessment using metaphase analysis or other techniques should be done. Loss of crypt cells and GI barriers may allow pathogenic and opportunistic bacterial infections. Use alimentary nutrition to encourage crypt cell growth. Avoid parenteral nutrition and central intravenous lines. Anticipate anaerobic colonization. All surgical procedures must be accomplished in initial 36 – 48 h after irradiation. Any additional surgery must be delayed until 6 weeks post-exposure.

5.0 – 8.0 Gy	Moderate to severe nausea and vomiting in 50 – 90% of personnel. Early: Moderate fatigability and weakness in 80 – 100% of personnel. Frequent diarrhea	ONSET: under 1 h END: indeterminate. May proceed directly to GI syndrome without a break	At 10 d – 5 weeks, 50 – 100% of personnel will develop pathogenic and opportunistic infections, bleeding, fever, loss of appetite, GI ulcerations, bloody diarrhea, nausea, severe fluid and electrolyte shifts, capillary leak, hypotension	Tertiary level intensive care required to improve survival. Fluid and electrolytes for GI losses. May require transfusion and/or colloids. Cytokines for immunocom-promised patients. Specific antibiotic therapy for infections, to include antifungals. Will require GI decontamination with quinolones. Use alimentary nutrition.	At low end of exposure range, death may occur at 6 weeks in more than 50%. At high end of exposure range, death may occur in 3 – 5 weeks in 90%.	Early evacuation to tertiary level medical center before onset of illness. Patients will require extensive reverse isolation to prevent cross contamination and nosocomial infection.	Practically no lymphocytes after 48 h. Severe drop in granulocytes and platelets later. In pure radiation exposure scenarios, these patients will require highest priority evacuation. The latent period between prodromal symptoms and manifest illness may be very short. When this radiation injury is combined with any significant physical trauma or infection, survival rates will approach zero. All surgical procedures must be accomplished in initial 36 – 48 h after irradiation. Any additional surgery must be delayed until 6 weeks post exposure. Partial marrow shielding may complicate bone marrow transplant. Steroid therapy is effective.

Dose	Initial Symptoms	Initial Symptoms Interval Onset - End	Medical Problems	Indicated Medical Treatment	Disposition without Medical Care	Disposition with Medical Care	Clinical Remarks
8.0 – 30+ Gy	Severe nausea, vomiting, fatigability, weakness, dizziness and disorientation. Moderate to severe fluid and electrolyte imbalance, hypotension, possible high fever, and sudden vascular collapse.	ONSET: less than 3 min END: death	Probable death at 2 – 3 weeks. Minimal if any break between prodromal syndrome and manifest illness. At high radiation levels, CNS symptoms predominate, with death secondary to cerebral vascular incompetence.	Supportive therapy in higher dosage ranges. Aggressive therapy if pure radiation injury and some evidence of response.	Expectant category	If assets are available, evacuate to a tertiary level medical center during manifest illness. Patients will require extensive reverse isolation to prevent cross contamination and nosocomial infection. Most patients will remain expectant.	Bone marrow totally depleted within days. Bone-marrow transplant may or may not improve ultimate outcome due to extensive intestinal damage and late radiation pneumonitis and fibrotic complications. Even minor wounds may prove ultimately fatal. Aggressive therapy is indicated when resources are available and transport to a tertiary care medical center is possible.

[a] This Table has been adapted from the *Medical Management of Radiological Casualties* (AFRRI, 1999).

Appendix B

Current Command and Control Policies and Structures

B.1 Purpose and Limitations

Any significant terrorist event involving the dispersal of radioactive materials will result in a large-scale response from multiple levels of government and several nongovernmental entities. One of the most formidable challenges facing the leadership will be how to organize and control these responding forces.

B.2 Domestic Response

B.2.1 *Local and State Governments*

Responsibility for public health and safety rests with local and state officials. While the organizations and capabilities vary from state to state, most local and state governments have evolved a common approach to emergency command and control.

In the 1970s, southern California firefighters developed a system to organize the many resources needed to fight wildfires. That system, the ICS, proved so effective that most local and state governments, with the assistance of FEMA, have now institutionalized ICS as the standard approach to command and control.

The ICS is essentially a "top down" approach. When an incident occurs, the senior first responder on the scene (normally a fire department official) becomes the IC and the command and control structure is expanded from that point forward based on the scope and duration of the incident. The initial IC remains in charge until and unless relieved by a more senior responder.

The ICS is organized on a functional basis. In addition to the Incident Command component, the ICS structure has four general staff sections called the planning, operations, logistics, and finance/administrative sections. The planning section collects, evaluates and disseminates information on the development of the incident and the status of resources. The planning section also creates the Incident Action Plan, which defines the response activities and the resources for a specified time period. The operations section directs tactical operations. The logistics section is responsible for facilities, services, materials and communications. The finance/administrative section tracks support, personnel, arrival time, costs, stay times, and other administrative matters. These sections each have their own manager, and can establish subsections to deal with different aspects of a critical event. In addition to the general staff positions of the ICS, the IC may appoint, if deemed necessary, three command staff positions to handle the media, safety and liaison. The command staff appointees report directly to the IC.[19]

In most states, the primary responsibility rests with local officials; the state is there to provide assistance and additional resources to the responding local authorities. When state resources arrive on scene, they are integrated into the existing ICS structure.

B.2.2 *Federal Domestic Response: Crisis Management*

Within the United States, primary responsibility for crisis management in the event of a terrorist incident rests with the federal government.[20] The DOJ, acting through the FBI, is the LFA (PDD, 1995/1998).

Upon receiving a threat or other information indicating a possible terrorist use of nuclear material, the FBI conducts a formal threat credibility assessment using experts from both FBI and DOE (FBI, 1998; page 9). Due to the large number of hoax threats and the incremental nature of incoming information, an assessment is conducted before deploying federal assets. The FBI has developed a four-tiered system to characterize the threat and describe the thresholds at which specific federal assets will be called.

[19]Additional information on the Incident Command System may be obtained from the National Fire Academy, U.S. Fire Administration, Emmitsburg, Maryland.

[20]See, for example, U.S. Code, Title 18, Section 2332b(f).

- **Threat Level 4**: *Minimal Threat*—the threat condition does not justify unusual actions and agencies continue to operate under normal day-to-day conditions.
- **Threat Level 3**: *Potential Threat*—there are indications of a threat, but the threat has not yet been assessed as credible. At this level, the FBI institutes the assessment process and begins to develop contingency deployment plans for follow-on resources.
- **Threat Level 2**: *Credible Threat*—the assessment has confirmed the involvement of a WMD and indicated that the threat is credible. The FBI would probably deploy a Domestic Emergency Support Team, a specialized United States government interagency team designed to provide expert advice on requirements and resources. At this point the primary focus remains on law enforcement actions designed to prevent and resolve the threat. However, selected assets will be pre-positioned and a JOC will be established to manage the developing crisis in the interagency environment incorporating law enforcement planning concerns with consequence-management concerns.
- **Threat Level 1**: *WMD Incident*—a WMD terrorism incident has occurred and the full federal response will be required. FEMA would lead the federal government's efforts to respond to the devastation through consequence management in support of the FBI as the LFA (FBI, 1998; page 12).

The FBI plan for command and control contemplates the appointment of a senior FBI official as the OSC.[21] The OSC will establish a JOC which, in addition to the Command Group, will have an Operations Group, a Support Group, and a Consequence Management Group. The Operations Group is responsible for intelligence, tactical operations, technical support, investigations, negotiations, and surveillance. The Support Group will provide support through the following components: media, administrative, liaison, legal and logistics. The Consequence Management Group will consist of representatives from other federal agencies such as FEMA, DOE, DOD, EPA, PHS, as well as local and

[21]Initially, the OSC will probably be the special agent in charge of the local FBI office. However, the FBI has designated a small number of specially trained SAC's as potential OSC's for WMD incidents. It is likely that one of these specially trained potential OSC's will be deployed to the scene to assume the duties of the OSC.

state representatives. The Consequence Management Group will address pre-release and post-release consequence operations (FBI, 1998; page 8).

The Command Group is the core of the JOC. It is composed of the FBI OSC and senior officials with decision-making authority from local, state and federal supporting agencies, as appropriate. The Command Group will jointly determine strategies, tactics and priorities. While the FBI OSC always retains authority to make federal crisis-management decisions, operational decisions will be made cooperatively to the greatest extent possible. If conflicts arise in priorities for allocation of critical federal resources between crisis management and consequence management, the FBI OSC and the senior FEMA official will provide, or obtain from higher authority, an immediate resolution.[22]

In addition to establishing the JOC, the OSC will establish an Interagency JIC to serve as a focal point for the coordination and provision of information to the public and the media concerning the response to the emergency (FBI, 1998; page 9).

A critical interface exists between the JOC and the ICS Incident Command Post (ICP). The JOC is a stand-alone unified federal command center whose function is to coordinate the federal crisis-management response to the incident, manage the investigation, and track local, state and federal consequence-management actions. The JOC, along with the FEMA Disaster Field Office, functions as the focal point for federal assistance to the local and state responders when the scope of the incident exceeds their needs. The JOC is not intended to replace the ICS ICP.

When an incident occurs and an ICP is established, the JOC will deploy a Forward Command Element (FCE) to the incident site to augment the resources of the ICP. The FCE will merge with the ICP as part of the unified command. The function of the FCE is to coordinate federal resources to assist the IC if needed and to insure the concerns of evidence and crime scene preservation are properly addressed. The JOC FCE will be task organized based on the incident and will include interagency representation if appropriate. The FBI commander will report to the OSC in the JOC, but will coordinate all FCE activities through the ICS unified command.[22]

Another critical interface is the transition from crisis management to consequence management. Depending upon the nature of the event, crisis management and consequence management may

[22]Command and Control Issues; FBI Scheme and the Incident Command System, undated (Federal Bureau of Investigation, Washington).

be occurring simultaneously. The FBI, however, will remain in charge until such time as the Attorney General transfers the LFA role to FEMA (FBI, 1998; page A-5).

B.2.3 *Federal Domestic Response: Consequence Management*

Primary responsibility for consequence management rests with local and state officials. The federal government is strictly in a supporting role (unless the incident is so serious as to disrupt the ability of the local and state governments to respond) (PDD, 1995/1998). While the response organizations and capabilities of the several states vary, each state has consequence-management capabilities and radiological expertise. However, in a significant incident, the local and state authorities will undoubtedly desire federal assistance.

Just as in the case of a natural disaster, FEMA as the LFA, is responsible for coordinating the federal response and is the conduit through which state officials can request federal assistance (PDD, 1995/1998). FEMA, in coordination with other federal departments, developed a Federal Response Plan (FRP) (FEMA, 1999). The plan, which is intended to apply to any emergency or disaster, defines 12 emergency support functions (ESF) and assigns primary responsibility for each ESF to a federal department or agency.

As a result of PDD-39 (PDD, 1995/1998), FEMA, with the assistance of DOD, DOE, HHS, FBI and EPA, published a Terrorism Incident Annex to the FRP. The Terrorism Incident Annex describes the relationship between the FRP, the Federal Radiological Emergency Response Plan (FRERP) (FEMA, 1996), and the National Oil and Hazardous Substances Pollution Contingency Plan (NCP) (EPA, 1994).

Paragraph V.D.2 of the Terrorism Incident Annex notes that the FBI has concluded formal agreements with potential lead federal agencies of the FRERP. If the FRERP is implemented concurrently with the FRP, the federal OSC under the FRERP will coordinate the FRERP response with the FEMA official who is responsible under PDD-39 for coordination of all federal support to local and state government (PDD, 1995/1998).

Paragraph V.D.4 of the Terrorism Incident Annex provides that, if the NCP is implemented, the Hazardous Materials On-Scene Coordinator under the NCP will coordinate, through the Hazardous Material Chair, the NCP response with the FEMA official who is responsible under PDD-39 for coordination of all federal support to local and state government (PDD, 1995/1998).

When the FBI determines that there is a credible threat, the FBI will notify FEMA. Based on the circumstances, FEMA will deploy representatives on the Domestic Emergency Support Team and may alert other federal agencies supporting consequence management. FEMA will determine the appropriate agencies to staff the JOC Consequence Management Group and advise the FBI. With FBI concurrence, FEMA notifies the agencies to request that they send representatives to the JOC. Representatives may be requested for the JOC Command Group, the JOC Consequence Management Group, and the JIC. FEMA may also activate an Emergency Support Team and convene an executive-level meeting of the Catastrophic Disaster Response Group. When appropriate, FEMA will consult with the affected Governor and the White House to determine whether federal assistance is required and if FEMA is permitted to use the authorities of the Robert T. Stafford Disaster Relief and Emergency Assistance Act (Stafford Act) (FEMA, 1999; page TI-7).

If a release occurs in the course of an incident or an incident occurs without warning, FEMA will consult immediately with the Governor and the White House to determine whether federal assistance is required and whether FEMA is permitted to use its authorities under the Stafford Act. During the consequence-management response, the Regional Operations Center Director or, if appointed, the Federal Coordinating Officer retains authority to make federal consequence-management decisions at all times. However, these decisions will be made cooperatively to the greatest extent possible. Resolution of conflicts between the crisis-management and consequence-management responses will be resolved by the FBI OSC and the senior FEMA official at the JOC or will be referred to higher authority (FEMA, 1999; page TI-8).

The FRP may be implemented concurrently with other Federal Emergency Operations Plans (FEMA, 1999; page 3). One such plan, the FRERP, was developed after the TMI accident and remains in effect (FEMA, 1996). Under the FRERP, DOE will establish a Federal Radiological Monitoring and Assessment Center (FRMAC) to deal with radiological issues. The FRMAC, composed of DOE and EPA personnel as well as local and state radiological experts, will characterize the contamination and provide health and safety advice to the responsible officials. During the intermediate phase of an incident, the leadership of the FRMAC passes to EPA, which is responsible for long-term monitoring and for providing advice to local and state officials on protection actions and for restoration during the recovery phase.

B.3 Overseas Response

Overseas, the nation in which the event occurs is responsible for both crisis management and consequence management. However, in cases in which there is no functioning, effective government, the organizations providing humanitarian relief may be forced to take charge of the situation.

The U.S. Department of State (DOS) is the LFA for United States' response to radiological terrorism occurring outside the country (PDD, 1995/1998). In a crisis, the DOS Operations Center would form a 24 h task force to coordinate the response. DOS would also lead an interagency Foreign Emergency Support Team (FEST) which, upon the advice of the Ambassador, would deploy to the affected embassy to provide expert advice. The FEST would be tailored to the type of incident, and the needs of the Ambassador and the host nation. In the context of a terrorist event involving radiological material, the FEST could provide guidance on terrorist policy and incident management, secure communications, and special equipment and expertise appropriate for the nuclear threat (GAO, 1997; page 50). The FEST may be deployed regardless of whether the host nation requests assistance because the embassy will have to deal with potentially affected United States personnel and commercial interests.

The United States government provides consequence-management assistance upon the request of the host government when the Ambassador determines that such assistance is necessary and in the best interests of the United States (GAO, 1997; page 64). If the determination is made to provide assistance, the DOS would include, as an integral part of the FEST, a Consequence Management Response Team (CMRT). The CMRT would be led by DOS and would include representatives from DOD,[23] the Office of Foreign Disaster Assistance,[24] and selected experts from other departments and agencies such as DOE and PHS. The CMRT would be the single focal point for consequence management and would develop, in coordination with the Country Team, the initial Consequence Management Plan.[25]

[23]The DOD element would include Humanitarian Assistance Survey Team liaison officers.

[24]The Office of Foreign Disaster Assistance would also deploy a Disaster Assessment Response Team.

[25]Minutes from February 20, 1998 interagency meeting hosted by DOS on the Consequence Management Response Team.

In a major incident, the consequence management and humanitarian assistance would be likely to include, in addition to host nation and United States resources, help from a number of international organizations, nongovernmental organizations, and private organizations.[26] The responding international organization would probably include the International Committee of the Red Cross and several elements of the United Nations.[27] Depending on the location of the incident, regional organizations such as the European Community Humanitarian Office and the North Atlantic Treaty Organization may be involved (Heddings, 1999; page 46).

While the assistance of the myriad of responding organizations will increase the resources available to the consequence managers, their presence will greatly complicate the command and control relationships. The CMRT will have the difficult task of coordinating the United States response with the host nation and these many other organizations.

If the incident occurs within the boundaries of one of the signatory states, the consequence-management assistance will be provided within the framework of an international convention developed in the aftermath of the Chernobyl accident (IAEA, 1987). This international convention protects the responders by holding them harmless from legal liability for any of their actions in support of the humanitarian assistance. However, if the host nation is not a signatory or refused to accede to this provision, there is the potential for criminal or civil liabilities. In order to protect United States responders in such cases, the United States government will probably insist upon a bilateral agreement with the host nation.

[26]The United Nations has recognized some 2,700 nongovernmental organizations with humanitarian missions and there are some 475 private organizations currently registered with the U.S. Agency for International Development (Heddings, 1999; page 48).

[27]Organic United Nations elements likely to respond include the Department of Humanitarian Affairs, High Commissioner for Refugees, Development Program, International Children's Emergency Fund, and the Environment Program (Heddings, 1999; page 42). Specialized agencies affiliated with the United Nations as provided for in Article 57 of the United Nations Charter will also be present. These include the International Atomic Energy Agency, the World Health Organization, the Food and Agriculture Organization, the World Food Program, World Bank and International Monetary Fund, World Meteorological Organization, Pan American Health Organization, and the International Telecommunications Union (Heddings, 1999; page 44).

B.4 Summary

This discussion of the federal response sounds complex, and integration of the responding resources will be a challenge. However, the basic concepts are relatively straightforward. Within the United States, the federal government is responsible for crisis management and the FBI is the LFA. Local and state authorities are responsible for consequence management and FEMA is the LFA. Overseas, the host nation is in charge of both crisis management and consequence management. DOS is the LFA and coordinates the United States response with the host nation and other responding organizations.

Appendix C

Current Federal Communications Policy and Plans

In order to deter terrorism, it is United States government policy to provide a clear public position that United States policies will not be affected by terrorist acts and that the United States will not allow terrorism to succeed. This counterterrorism policy has been established most recently in Presidential Decision Directive 39. In the United States, federal consequence-management response to terrorism is led by FEMA or, if large-scale casualties and infrastructure damage occur, the President may appoint a personal representative. The FRP, its Terrorism Incident Annex, and the FRERP (FEMA, 1996) guide consequence-management activities in response to domestic terrorist attacks. The plans include provisions for public information activities to ensure the coordinated, timely and accurate release of information to the news media and to the public about disaster-related activities.

Federal emergency information plans and policies are predicated on the assumption that no single agency at the local, state, federal or private-sector level has the authority or expertise to act unilaterally in response to threats or acts of terrorism involving WMD (FEMA, 1999).

The FBI and FEMA work together to establish and operate a JIC in the field which is the focal point for information to the public and the media concerning the federal response to the emergency. However, local and state governments exercise primary authority to respond to terrorism consequences, the federal government provides assistance, as required. FEMA plans designate that a state public affairs officer jointly manages the JIC with FEMA's lead public affairs officer. Local and state consequence-management public information personnel should join federal public information personnel in the JIC. FEMA public information plans are published in its FRP for disaster response, the Terrorism Incident Annex, and the FRERP. Immediately following the incident, communications with the public involve providing warnings and disseminating protective action and emergency information.

If there is no terrorist act, or if a terrorist act does not result in major consequences, FRP structures, such as the FEMA/state-lead JIC, disengage at appropriate times. After disengagement, operations by individual or multiple federal agencies under other federal plans may continue to support local and state governments with long-term efforts such as hazard monitoring, environmental decontamination, and site remediation.

The FEMA emergency information objectives are to:

- Instill confidence in the community that all levels of government are working in partnership to restore essential services and help individuals begin to put their lives back together;
- Work with media to promote a positive understanding of federal and state response, recovery and mitigation programs;
- Provide all target markets with equal access to timely and accurate information about disaster response, recovery and mitigation programs;
- Manage expectations so that disaster victims have a clear understanding of all disaster response, recovery and mitigation services available to them; and
- Support local and state efforts to reach disaster victims with specific program information (FEMA, 1998).

Information release authority. During crisis management, the FBI is the federal government release authority. In consequence-management situations, FEMA is the federal government release authority. The JIC, in coordination with appropriate response organizations, releases information.

Information coordination. Coordination ensures information is accurate, and intelligence and law enforcement resources and efforts are not compromised. Information intended for the news media and the public is coordinated with the Federal Coordinating Officer, other federal organizations, and state and local officials.

Federal government communication responsibilities. Federal plans call for the JIC to be established by the LFA (FBI crisis,

FEMA consequence management). However, it is understood that local responders usually arrive on-scene first and initially form a JIC. FEMA's public information officer does not exercise editorial or policy control over other agencies' release of information about their own policies, procedures or programs.

State communication responsibilities. State governments are an essential element of the FBI and FEMA public information response effort. They are expected to participate in and share JIC resources. FEMA plans designate a state public information officer to jointly manage the JIC with FEMA's lead public information officer.

Local communication responsibilities. Local government response organizations are key to FBI and FEMA public information efforts. They are expected to participate in and share JIC resources.

Communication plan. FEMA's emergency information response plans are outlined in its Emergency Information Field Guide and its Community Relations Operations Manual. These documents contain extensive checklists and functions for public information operations.

Release of cleared information. This information will be made available through various means to include: press releases, briefings, interviews, special press handouts, and the FEMA Internet web site.

Media communication methods:

- emergency broadcast system/FEMA Recovery Channel/ FEMA Recovery radio network
- public service announcements
- Internet (video stream)
- homepage
- E-mail
- information hotlines
- site access

Public communication methods:

- emergency broadcast system/FEMA Recovery Channel
- radio
- public service announcements
- Internet homepage/FEMA web site

- E-mail
- media
- FEMA Recovery Times newsletter
- information hotlines/FEMA helpline
- public meetings
- advertisements

Community outreach. The Federal Coordination Officer community outreach plans call for close coordination with affected state and local groups to assist in rapid dissemination of information and to identify public concerns about the consequence-management response and to take appropriate action.

Appendix D

Sample Joint Information Center Checklist[28]

- Notification of possible terrorist event/confirmation of terrorist event.
- Ensure OSC/lead official knows you.
- Communicate with appropriate federal, state, local, and voluntary organizations/public information personnel.
- Establish permanent public information liaison(s) in the operations center and other appropriate response elements.
- Request, as appropriate, additional public information personnel through OSC/lead official.
- Advise the OSC/lead official.
- Establish the JIC.
- Ensure adequate communication, transportation, logistic, computer/information system, graphics, audiovisual, and administrative support for the JIC. Have/obtain contract support or local purchase authority.
- Establish coordination procedures: prepare, authorize and disseminate information on the terrorist incident and its effects in coordination with the OSC/lead official (or his designee) and appropriate local, state, federal, technical, security and legal personnel.
- Develop/review/implement the public information plan and daily key messages. Ensure plans include public information activities related to evacuation and/or sheltering, public safety announcements, and the worried well.
- Identify technical liaison personnel.

[28]Extensive checklists are published in the FEMA Emergency Information Field Guide (FEMA, 1998) and the FEMA Operations Guide, Community Relations in Federal Disaster Operations (FEMA, 1999).

- Establish a media briefing center.
- Prepare briefers and interviewees.
- Coordinate and conduct media briefings.
- Coordinate media interviews.
- Analyze news, provide feedback to JIC staff and response organizations.
- Ensure appropriate access and transportation, communication and logistic support for the media.
- Ensure incoming personnel are briefed on the situation, JIC organization, policies, procedures and types of nonreleasable information. Make released material available for JIC staff review.
- Ensure responding public information personnel are assigned specific tasks.
- Develop and schedule shifts.
- Log events, queries and responses.
- Prepare situation reports.
- Establish and participate in a public outreach program.
- Ensure establishment of an 800 number for the public to call.
- Establish an internal information program.
- Review local area networks, status boards, attend meetings and situation updates as well as shift turnovers.

Appendix E

Sample Pre-Prepared Public Information Statements

In many cases, information can be prepared in advance to provide background details and fill-in-the-blank statements that can be revised, as appropriate, and carefully coordinated for relevancy and possible release during an actual event. All pre-drafted information regarding the status of the disaster must be verified before release.

E.1 Sample Public Safety Statements [29]

The following statement is for use when there is a possibility of radiological exposure to the public. Local fire personnel or police will normally release it.

> In the interest of public safety and law enforcement requirements, the area around the incident site is being monitored and a barrier (is being)/(has been) established around it. Radioactive material may have been released, so there is a possibility of radiation exposure in the restricted area. This area is also a crime scene.
>
> It is important that the movement of people into and out of the restricted area be strictly controlled. For the time being, only members of the emergency services, local, state and federal response forces are being allowed inside the area. The public should stay away to reduce the

[29] Adapted from DOD Directive 5230.16.

possibility of radiation exposure from this incident and to facilitate response efforts.

The following statement is for use when there is a possibility of radiation exposure to the public and sheltering/evacuation is recommended.

In the interest of public safety and law enforcement requirements, the area around the incident site is being monitored and a barrier (is being)/(has been) established around it. Radioactive material may have been released, so there is a possibility of radiation exposure. This area is also a crime scene. The highest levels of contamination are expected to be there. However, radioactive material may have been carried downwind beyond the established perimeter of the restricted area.

As a precaution, the public is advised to [take shelter in (location)]/[evacuate the following areas...]. We will continue to monitor the site to determine whether there could be (any risk)/(any further risk) to the public.

It is important that the movement of people into and out of the restricted area is strictly controlled. Only members of the emergency services, local, state and federal response forces are being allowed inside the area. The public should stay away to reduce the possibility of radiation exposure from this incident and to facilitate response efforts.

The following statement is for use when radioactive release has been confirmed.

A release of radioactive material has been detected. The highest levels of contamination are expected to be within the restricted area, which is also a crime scene. However, radioactive material may have been carried downwind beyond the perimeter of the restricted area.

As a precaution, the public is advised to [take shelter in (location)]/[evacuate the following areas...]. We will continue to monitor the area to establish the extent of contamination and determine the risk to the public.

It is important that the movement of people into and out of the cordoned area is strictly controlled. Only members of the emergency services, local, state and federal

> response forces are being allowed inside the area. The public should stay away to reduce the possibility of radiation exposure from this incident and to facilitate response efforts.

Local emergency response personnel will normally issue public safety statements advising precautions to be taken against potential exposure to radiological material. For cases in which the IC has determined that there has been a release of significant amounts of radioactive materials, the following information should be released to persons in affected areas as soon as possible after the incident.

> Until the amount of radiological contamination is determined, the following precautionary measures are recommended to minimize risk to the public:
>
> - remain inside and minimize opening doors and windows.
> - children should not play outdoors
> - fruit and vegetables grown in the area should not be eaten
> - turn off fans, air conditioners, and forced air heating units that bring in fresh air from the outside. Use them only to recirculate air already in the building
>
> The inhalation of radioactive material (plutonium, uranium) is not an immediate medical emergency.
>
> Trained monitoring teams will be moving through the area wearing special protective clothing and equipment to determine the extent of possible radiological contamination. The dress of these teams should not be interpreted as indicating any special risk to those indoors. If you are outside, proceed to the nearest permanent structure. If you must go outside for critical or lifesaving activities, cover your nose and mouth and avoid stirring up and breathing any dust. It is important to remember that your movement outside could cause you greater exposure and possibly spread contamination to those already protected.
>
> Local, state and federal personnel are responding to the (terrorist)/(potential terrorist) attack.
>
> In the interest of public safety and to assist emergency response teams, authorities request that individuals

within the vicinity (define) stay inside, with doors and windows closed, unless advised to do otherwise by the police.

Further statements will be made when there is more information. Please listen for announcements on local radio/television (name stations and frequencies). Check the Internet at (web site).

E.2 Potential Key Messages

During a disaster response, there are important messages that emergency responders may take for granted and are often overlooked. Nevertheless, they are important themes that need to be emphasized. Examples are provided below.

Safety

- Public health and safety is our first priority.
- Trained local, state and federal civilian and military personnel are responding.
- Please stay away from the restricted area to ensure your safety and so we can work efficiently.
- Preventing further injuries or loss of life is paramount.

Sympathy

- We deeply regret this incident has occurred.
- Our thoughts and condolences go out to families and friends of the victims.

Cooperation

- We are working closely with all involved local, state and federal organizations.
- Experienced federal and civilian specialists with the most advanced equipment are responding to this emergency.
- We want to understand your concerns.

Disclosure

- We are here to coordinate the initial consequence-management response.

- We will give you information as soon as it becomes available for public release.
- We want to answer your questions.
- We do not have all the answers.

Disaster assistance

- Procedures will be established to handle requests for disaster assistance.

Radiation and health risk

- Radiation exposure can have short- and long-term consequences to human health.

Health effects depend on the radiation dose received and many other factors including length of time exposed, distance from the radiation source and protection such as shelter or clothing worn at the time of exposure. Therefore, an individual's health risk from radiation exposure from this incident may be uncertain.

- Children exposed to radiation can be more at risk than adults.
- Radiation exposure, like exposure to the sun, is cumulative.
- Exposure to radiation may cause cancer in the long term. Exposure to a very high dose of radiation can cause death in the short term.
- If someone exposed to radiation from this incident eventually develops cancer, medical and scientific personnel will not be certain that this exposure caused the cancer.
- There is no more effective or necessary screening for cancer than existing medical methods (mammograms, pap smears, colon cancer tests). People potentially at risk of developing cancer in the future due to radiation exposure resulting from this incident should see their physicians for an annual physical.

Much is known about:

- how to minimize human exposure to radiation
- how to treat people exposed to radiation
- how to decontaminate people exposed to radiation
- how to decontaminate animals exposed to radiation

- how to cleanup property contaminated with radioactive materials

Changing and varying information:

The process of providing information is complicated by the complex nature of health effects associated with radiation. It is also complicated because the information is based upon estimation techniques and scientific monitoring of radiation levels that will vary in different locations and will change over time due to weather conditions and other variables.

E.3 Potential Consequence-Management Questions/Answers

Those responsible for public information can anticipate certain types of questions involving technical information as a result of a radiological disaster. Many of these questions are provided below. When possible, suggested answers are also presented. Some of the recommended answers (*e.g.*, definitions of radiation quantities and units) are intended for public release and should not be used for rigorous scientific purposes.

Assistance

Q: What can volunteers do to assist?
Q: Where can volunteers go to assist?

Casualties

Q: How many deaths/injuries were there?
Q: What caused the deaths/injuries (explosion, radiation)?
Q: Have responders been injured or killed? If so, how?
Q: To which hospital(s) will the injured be transported?
Q: Does/do the hospital(s) have staff to monitor contaminated patients?
Q: Will the hospital staff and patients be in danger from treating/being near the contaminated patients?

Claims

Q: Who will pay for the loss and damage?
Q: Where can people file claims?
Q: How soon is financial assistance available?

Domestic animals/wildlife

Q: What is being done to protect pets, livestock and wildlife?
Q: What should I do if I suspect my pet has been exposed to radiation?

Environment

Q: What is the effect on the water system?
Q: What is the effect on well water?
Q: What is the effect on nearby rivers/lakes/streams?
Q: Was there property damage? If so, what's the estimated cost?

Hazards

Q: What are health officials' most immediate public health concerns?

Q: What is an external radiation hazard?
A: An external radiation hazard can result when a source of radiation, for example, a quantity of radioactive material, is outside of (external to) the body. Time, distance and shielding protect the body from external radiation.

Q: What is an internal radiation hazard?
A: An internal radiation hazard can result from the deposition of radioactive material inside the human body. This can occur as a result of a person breathing radioactive material present as a dust, vapor or gas; through the ingestion of radioactive materials either in solid or liquid form; through the intake of radioactive materials through cuts or wounds; or through absorption of radioactive materials through the skin.

Q: How is radiation measured?
A: There are several ways to measure radiation. We can measure the actual energy in the air, or the energy absorbed or released by a substance. Ground surveys and aerial measurements can be made to determine the extent of contamination on the ground. Air monitoring stations

can be set up to detect contamination in the air. Sampling and testing of soil, vegetation and crops can be conducted to determine the amount of contamination present.

Q: How can people tell if they have been contaminated?

A: If people think they were exposed to radioactive material from the terrorist event, they should contact appropriate local, state or federal authorities and arrange to be monitored. Survey instruments can be used to assess possible contamination of their clothing with radioactive materials. Urine and/or fecal samples can be collected and analyzed to estimate the quantities of radionuclides taken into their bodies. The amounts of radioactive materials deposited in their lungs can be estimated by counting the gamma radiation emerging from their chest wall, through a so-called "lung count."

Q: What is a lung count? Whole-body count?

A: Many of the more penetrating gamma radiations emitted by radionuclides within a person's body will not be absorbed. That is to say, they will escape from the body. A whole-body count is a procedure in which these escaping gamma radiations are counted. The total count provides an indication of the amount of radioactive material in the body. In the case of a whole-body count, all the gamma radiations escaping from the body are counted. In the case of a lung count, only those emerging from the chest wall are counted. Spectroscopic analyses of the energies of the gamma radiations can be used to identify the specific radionuclides present in the body.

Q: Can people die from radiation exposure in hours, days, weeks, months or years from now? What are the short- and long-term effects of radiation on people?

A: Massive exposure to radiation can cause death within a few hours or days. Smaller doses can cause burns, nausea, loss of hair, loss of fertility, and pronounced changes in the blood. Even smaller doses, too small to cause any immediate visible damage, are thought to increase the probability of developing cancer or leukemia, congenital abnormalities in children exposed *in utero* including physical deformities, diseases, and mental retardation.

Q: Is a child's exposure to radiation from this incident more hazardous than an adult's exposure?

A: Based on studies of exposed populations, children have a slightly higher risk of cancer following exposure to ionizing radiation.

Q: What precautions should residents take to avoid exposure/further exposure?

A: Protection:

- respiratory protection (includes closing windows and doors)
- protective clothing
- cover open cuts and wounds
- washing/decontamination
- food controls

Q: What are the United States government standards for radiation exposure?

A: The United States government has set the maximum acceptable levels for occupational exposure to radiation at 5 rem (5,000 mrem or 50 mSv) y^{-1} from all human-made sources combined and has set a variety of lower limits for protection of the general public that depend on the source of radiation.

Q: How does radiation hurt people?

A: Radiation can damage genetic natural (DNA) in the body's cells, especially dividing cells. If a small amount of radiation is absorbed by the body, it does not always damage the cells. If it does, the cells can sometimes repair themselves. Damaged cells can die right away, or if they survive, may be transformed into cells that could cause a tumor.

Q: How much radiation can cause cancer?

A: No one is sure how much radiation can cause cancer but we assume that the risk of cancer is proportional to the absorbed dose. Low doses could cause cancers 5 to 30 y or longer after exposure. However, it is important to remember that people are exposed to radiation every day from a variety of sources in the natural environment. The amount of radiation that is absorbed by the body is quantified with a unit called a sievert (Sv) or a millisievert (1 mSv, one-thousandth of a sievert). Exposure to background radiation, from sources such as radon gas, outer space, rocks and soil, results in the body absorbing about 0.05 mSv each week. Normally, 200 out of 1,000 United States citizens would be expected to die from cancer. With what we know now, a dose of 50 mSv is thought to increase cancer deaths to 203 out of 1,000.

Q: How can doctors tell how much radiation people have been exposed to?

A: Medical personnel can screen people using biological dosimetry. Techniques can also be used to determine if individuals have radioactive materials in or on their bodies. These methods are used to determine how best to treat patients with radiological injuries.

Q: Is it safe to eat food and drink milk and water? What about eggs, fruit, livestock, fish and crops?

A: (Answer to be determined based on command and control guidance.) Fruit exposed to any residue cloud from the event may have contamination on its surface. Wash and peel fruit before eating.

Q: Are plants in gardens and agricultural produce in the area contaminated by radioactive material?

Radiation

Q: How much radioactive material has been released?

Q: What areas are radiologically contaminated and at what levels?

Q: Could the radiological contamination spread further?

Q: What radiological materials were involved?

Q: What is the highest radiation level and where is it?

Q: What is plutonium, and how can it harm people?

A: Plutonium is an artificially produced radioactive material. This radioactive element decays by emitting alpha particles and has a very long half-life. The range in air for alpha particles is only a few inches. This means that alpha radiation is not a hazard to people as long as it remains external to the body. Inhalation of airborne plutonium is normally the most hazardous exposure pathway. Following deposition in the lungs, it is transferred primarily to the liver and the bones from where it is cleared only very slowly.

Ingestion is normally a less hazardous pathway because plutonium is only minimally absorbed into the body as it passes through the gastrointestinal tract.

Q: What is uranium?

A: Uranium is a natural substance widely distributed over the earth. Uranium slowly reacts chemically (oxidizes) when exposed to air. In the air, the metal becomes coated with a layer of oxide that will make it appear from a golden-yellow color to almost black. It is an element having several radioactive isotopes with very long half-lives. Uranium also produces more than a dozen other radioactive substances as by-products, including radon gas. Tiny amounts of uranium are found almost everywhere on earth. Concentrated deposits are found in just a few places, usually in hard rock or sandstone, normally buried by earth and vegetation. Uranium has been mined in the southwest United States, Australia, parts of Europe, Russia, Namibia, South Africa, and Niger. At high concentrations, uranium is chemically toxic to the kidney. The radioactive by-products from the decay of uranium are generally more of radiological hazard than uranium itself.

Q: What is cobalt?

A: Stable cobalt is mined and is used in a variety of industrial applications. Radioactive cobalt is generally obtained from the irradiation of other metals in a reactor. Reactor-produced ^{60}Co is widely used as a source of radiation in industry and is widely used in medicine to treat cancer. Cobalt-60 has a 5 y half-life and emits beta radiation and penetrating gamma radiation.

Q: What is cesium?

A: Cesium-137 has a 30 y half-life and is one of the radioactive fission products created within a nuclear reactor during its operation. It can be absorbed into the food chain and can be an external and internal hazard. Cesium-137 sources are used to measure the thickness or density of material and for gamma radiography.

Q: What is radioactivity?

A: Radioactivity is the spontaneous emission of radiation from the nucleus of an unstable isotope.

Q: What is radiation?

A: Radiation is a form of energy. Although the term "radiation" is broad enough to include energy in the form of things such as sunlight, heat, radio waves and microwaves, the term is most commonly used to mean "ionizing" radiation. "Ionizing" radiation is that which is capable of removing orbital (or bound) electrons from atoms, thus producing electrically charged atoms/ions). There are three main types of radiation: alpha particles, beta particles, and gamma rays.

Alpha particles have a very short range and are easily shielded by a single sheet of paper. Alpha particles cannot penetrate the outer layers of skin and are not an external hazard. Radioactive material that emits them are an internal hazard if ingested or inhaled.

Beta particles have a longer range and are less easily shielded. Aluminum foil or glass will stop beta particles. They can penetrate the outer layer of skin and are an external and an internal hazard.

Gamma radiation has a very long range and is very difficult to shield. Unlike alpha or beta particles, gamma rays are electromagnetic energy waves (radio waves with a much shorter wavelength) similar to x rays. Concrete, lead or steel is needed to shield sources of gamma rays. The radiation can penetrate through the whole body; it is an external and an internal hazard.

Q: What is a becquerel?

A: A becquerel (Bq) is the special name of a unit by which the quantity of radioactive material is described. One becquerel is equal to one disintegration of a radioactive atom within a mass of radioactive material per second. Another unit often used to describe the quantity of radioactive material is the curie (Ci). One curie is equal to 37 billion becquerels.

Q: What is half-life?

A: The activity of radioactive material decreases with time. The half-life equals the period in which the activity decreases by half due to radioactive decay. Different radionuclides have different half-lives, from a fraction of a second to millions of years or more.

Q: What is contamination?

A: Contamination is the deposition and/or absorption of finely divided radioactive material, biological or chemical agents, or hazardous materials, structures, areas, people or objects.

Remediation

Q: What measures can be taken to cleanup the area now and in the future?

A: Response organizations, local, state and federal, will prepare a site remediation, or cleanup, plan. The process is lengthy and depends on the type of contamination and the site contaminated. There are temporary measures that can be taken to fix radioactive materials in place and stop the spread of contamination. These include "fixative" sprays such as flour and water mixtures, road oil, or water that can be used to wet ground surfaces and prevent resuspension.

Q: How much will it cost?

A: The cost will depend on the extent of contamination, site remediation methods, and the cleanup plan selected with community involvement.

Q: How long will cleanup take?

A: Site remediation is a lengthy process that can take years to complete, depending on the type of contamination, the site contaminated, and the remediation plan selected.

Q: How can I tell if my house is contaminated?

A: If your house was downwind from the residue cloud, it may be contaminated. You should contact local, state or federal authorities. They will arrange for a team to survey your house to detect possible contamination.

Response

Q: Who is in charge?

A: (Name of LFA) is responsible for coordinating the joint response. The senior local official is (name, organization). The senior state official is (name, organization). The senior federal official is (name, organization).

Q: When did authorities know about this incident?

Q: What response agencies are involved? What are their missions? What expertise do their response teams have? How many people are on the teams? What equipment do they bring? Where do they come from? How long does it take teams to respond following notification?

Q: How experienced are responders? How often do these organizations practice responding to similar radiological incidents?

Q: Under what authority does the LFA respond?

A: Presidential Decision Directive 39 (PDD-39), U.S. Policy on Counterterrorism, establishes policy to respond to terrorism directed against Americans. Responding federal agencies do so in accordance with the FRP (for Public Law 93-288, as amended) and its Terrorism Incident Annex.

Q: Who pays for response teams?

A: Responding organizations pay for all their expenses unless otherwise directed by the President.

Q: How many people are responding?

Q: Can I obtain copies of response plans?

A: The FRP is located on the Internet at www.fema.gov.

Sheltering/evacuation

Q: What areas are recommended for sheltering?

Q: What areas are recommended for evacuation?

Q: Why was sheltering recommended/ordered?

Q: Why wasn't sheltering recommended/ordered?

Q: How many people are/were affected by the sheltering order/recommendation?

Q: When will sheltering/evacuation guidance be lifted?

Q: When will residents be able to return home?

Q: When will businesses be able to reopen?

Appendix F

Federal and State Resources for Emergency Response and Planning Assistance

The following is a partial list of assets currently available within the United States.

F.1 U.S. Department of Energy Assets

The following are various organizations currently supported by DOE and that may be able to provide assistance in the event of a radiological emergency.

ACCIDENT RESPONSE GROUP (ARG)
Albuquerque Operations Office
PO Box 5400
Albuquerque, NM 87115
Telephone: (505) 845-4667 (24 h)
Specialty:

Primary accident response element for events or accidents involving nuclear weapons. Trained in weapon recovery and in evaluating, collecting, handling and mitigating radioactive and other weapons-associated hazards.

AERIAL MEASURING SYSTEM (AMS)
Nevada Operations Office
PO Box 98518
Las Vegas, NV 89193-8518
Telephone: (702) 295-1381 (24 h)

Specialty:

Aerial detection system for measuring extremely low levels of gamma radiation and locating and tracking airborne radiation. The system, also includes aerial photography and multi-spectral scanning capabilities.

ATMOSPHERIC RELEASE ADVISORY CAPABILITY (ARAC)

Oakland Operations Office
1301 Clay Street, Suite 865-N
Oakland, CA 94612
Telephone: (510) 637-1794 (24 h)

Specialty:

Computer-based, emergency response, and preparedness system that provides rapid predictions of the transport, diffusion and deposition of radionuclides or other toxic materials released into the atmosphere and dose projections to people and the environment.

FEDERAL RADIOLOGICAL MONITORING AND ASSESSMENT CENTER (FRMAC)

Nevada Operations Office
PO Box 98518
Las Vegas, NV 89193-8518
Telephone: (702) 295-1381 (24 h)

Specialty:

Temporary facility for production of compiled, quality-controlled monitoring and assessment data for the LFA and the local, tribal or state authorities involved in a radiological event.

NUCLEAR EMERGENCY SEARCH TEAM (NEST)

Nevada Operations Office
PO Box 98518
Las Vegas, NV 89193-8519
Telephone: (702) 295-1381 (24 h)

Specialty:

Provide technical assistance to FBI. The technical assistance includes such support as locating nuclear materials or devices that may be lost, stolen or associated with bomb threats.

RADIATION EMERGENCY ASSISTANCE CENTER/TRAINING SITE (REAC/TS)

Oak Ridge Operations Office
PO Box 2001
Oak Ridge, TN 37831
Telephone: (865) 576-1005
Specialty:

Direct or consultive assistance regarding medical and health physics problems associated with radiation accidents. Training in medical management for radiation accidents.

RADIOLOGICAL ASSISTANCE PROGRAM (RAP)

Region 1
Brookhaven Area Office
Upton, Long Island, NY 11978
Telephone: (516) 282-2200

Region 2
Oak Ridge Operations Office
PO Box 2001
Oak Ridge, TN 37831
Telephone: (865) 576-1005 (24 h)

Region 3
Savannah River Operations Office
PO Box A
Aiken, SC 29802
Telephone: (803) 725-3333 (24 h)

Region 4
Albuquerque Operations Office
PO Box 5400
Albuquerque, NM 87115
Telephone: (505) 845-4667 (24 h)

Region 5
Chicago Operations Office
9800 South Cass Avenue
Argonne, IL 60439
Telephone: (708) 252-5731

Region 6
Idaho Operations Office
785 DOE Place
Idaho Falls, ID 83402
Telephone: (208) 526-1515 (24 h)

Region 7
Oakland Operations Office
1333 Broadway
Oakland, CA 94612
Telephone: (510) 273-4237

Region 8
Richland Operations Office
Telephone: (509) 373-3800 (24 h)

Specialty:

Provides radiological assistance to other federal agencies, state, tribal and local governments, and Nuclear Regulatory Commission (NRC) licensees requesting assistance for events involving radioactive materials.

F.2 Information on Federal Emergency Management Agency Regions

REGION I (Boston, Massachusetts)
Serving: Connecticut, Massachusetts, Maine, New Hampshire, Rhode Island, Vermont
Federal Emergency Management Agency
JW McCormack Post Office and Courthouse Building
Room 442
Boston, MA 02109-4595
Telephone: (617) 223-4742
Fax: (617) 223-9567

REGION II (New York, New York)
Serving: New Jersey, New York, Puerto Rico, Virgin Islands
Federal Emergency Management Agency
26 Federal Plaza, Suite 1337
New York, NY 10278
Telephone: (212) 225-7204
Fax: (212) 225-7733

REGION III (Philadelphia, Pennsylvania)
Serving: Delaware, Maryland, Pennsylvania, Virginia, West Virginia, District of Columbia

Federal Emergency Management Agency
One Independence, 6th Floor
615 Chestnut Street
Philadelphia, PA 19106-4404
Telephone: (215) 931-5576
Fax: (215) 931-5539

REGION IV (Atlanta, Georgia)
Serving: Kentucky, Tennessee, North Carolina, South Carolina, Georgia, Florida, Alabama, Mississippi

Federal Emergency Management Agency
3003 Chamblee Tucker Road
Atlanta, GA 30341
Telephone: (770) 220-5466
Fax: (770) 220-5275

REGION V (Chicago, Illinois)
Serving: Illinois, Indiana, Michigan, Minnesota, Ohio, Wisconsin

Federal Emergency Management Agency
175 W. Jackson Boulevard, 4th Floor
Chicago, IL 60604-2698
Telephone: (312) 408-5528
Fax: (312) 408-5222

REGION VI (Denton, Texas)
Serving: Arkansas, Louisiana, New Mexico, Oklahoma, Texas

Federal Emergency Management Agency
Federal Regional Center
800 N. LOOP 288
Denton, TX 76201-3698
Telephone: (940) 898-5240
Fax: (940) 898-5121

REGION VII (Kansas City, Missouri)
Serving: Iowa, Kansas, Missouri, Nebraska

Federal Emergency Management Agency
2323 Grand Boulevard, Suite 900
Kansas City, MO 64108-2670
Telephone: (816) 283-7021
Fax: (816) 283-7098

REGION VIII (Denver, Colorado)
Serving: Colorado, Montana, North Dakota, South Dakota, Utah, Wyoming

Federal Emergency Management Agency
Denver Federal Center
Building 710, Box 25267
Denver, CO 80255-0267
Telephone: (303) 235-4812
Fax: (303) 235-4976

REGION IX (San Francisco, California)
Serving: Arizona, California, Hawaii and Nevada; and the Territory of American Samoa, the Territory of Guam, the Commonwealth of the Northern Mariana Islands, the Republic of the Marshall Islands, the Federated States of Micronesia, the Republic of Palau

Federal Emergency Management Agency
Building 105, PO Box 2998
Presidio of San Francisco
San Francisco, CA 94129-1250
Telephone: (415) 923-7277
Fax: (415) 923-7214

REGION X (Bothell, Washington)
Serving: Alaska, Idaho, Oregon, Washington

Federal Emergency Management Agency
Federal Regional Center
130 228th Street, SW
Bothell, WA 98021-9796
Telephone: (425) 487-4743
Fax: (425) 487-4777

F.3 Other Federal Assets

MEDICAL RADIOBIOLOGY ADVISORY TEAM (MRAT)

Armed Forces Radiobiology Research Institute
National Naval Medical Center
8901 Wisconsin Avenue, Building 42
Bethesda, MD 20889-5603
Telephone: (301) 295-0316 (24 h)

Specialty:

Consultation regarding medical and health physics problems associated with radiation accidents. Biological dosimetry capability and training program for medical personnel.

OFFICE FOR STATE AND LOCAL DOMESTIC PREPAREDNESS SUPPORT
810 Seventh Street, NW
Washington, DC 20531
Telephone: (202) 305-9887
Specialty:

Financial assistance for state and local governments to buy equipment, training, technical assistance, assessment of readiness, and emergency response exercises.

F.4 State Emergency Management Directors

Director
Alabama Emergency Management Agency
5898 County Road 41, PO Drawer 2160
Clanton, AL 35046-2160
Telephone: (205) 280-2285
Fax: 205-280-2444
Internet: http://www.aema.state.al.us/

Director
Alaska Division of Emergency Services
PO Box 5750
Fort Richardson, AK 99505-5750
Telephone: (907) 428-7039
Fax: (907) 428-7009
Internet: http://www.ak-prepared.com

Director
Arizona Division of Emergency Services
5636 East McDowell Road
Phoenix, AZ 85008
Telephone: (602) 231-6245
Fax: (602) 231-6356
Internet: http://www.state.az.us/es/

Director
Arkansas Department of Emergency Management
PO Box 758
Conway, AR 72033
Telephone: (501) 730-9750
Fax: (501) 730-9754
Internet: http://www.adem.state.ar.us/

Director
California Office of Emergency Services
2800 Meadowview Road
Sacramento, CA 95832
Telephone: (916) 262-1816
Fax: (916) 262-1677
Internet: http://www.oes.ca.gov/

Director
Colorado Office of Emergency Management
Division of Local Government
Department of Local Affairs
15075 South Golden Road
Golden, CO 80401-3979
Telephone: (303) 273-1622
Fax: (303) 273-1795
Internet: www.dlg.oem2.state.co.us/oem/oemindex.htm

Director
Connecticut Office of Emergency Management
Department of Public Safety
360 Broad Street
Hartford, CT 06105
Telephone: (203) 566-4343
Fax: (203) 247-0664
Internet: http://www.state.ct.us/dps/DFEBS/OEM/
entrance.htm

Director
Delaware Emergency Management Agency
165 Brick Store Landing Road
Smyrna, DE 19977
Telephone: (302) 659-3362
Fax: 659-6855
Internet: http://www.state.de.us/dema/index.htm

Director
D.C. Office of Emergency Preparedness
2000 14th Street, NW, 8th Floor
Washington, DC 20009
Telephone: (202) 727-6161
Fax: (202) 673-2290
Internet: http://www.fema.gov/dc-oep/

Director
State of Florida Division of Emergency Management
2555 Shumard Oak Boulevard
Tallahassee, FL 32399
Telephone: (904) 413-9969
Fax: (904) 488-1016
Internet: www.floridadisaster.org

Director
Georgia Emergency Management Agency
PO Box 18055
Atlanta, GA 30316-0055
Telephone: (404) 624-7000
Fax: (404) 635-7205
Internet: http://www.State.Ga.US/GEMA/

Director
Hawaii State Civil Defense
3949 Diamond Head Road
Honolulu, HI 96816-4495
Telephone: (808) 734-2161
Fax: (808) 733-4287
Internet: http://scd.state.hi.us

Director
Idaho Bureau of Disaster Services
4040 Guard Street, Building 600
Boise, ID 83705-5004
Telephone: (208) 334-3460
Fax: (208) 334-2322
Internet: http://www.state.id.us/bds/bds.html

Director
Illinois Emergency Management Agency
110 East Adams Street
Springfield, IL 62701
Telephone: (217) 782-2700
Fax: (217) 524-7967
Internet: http://www.state.il.us/iema

Director
Indiana Emergency Management Agency and Department of Fire and Building Services
302 West Washington Street, Room E-208
Indianapolis, IN 46204-2760
Telephone: (317) 232-3980
Fax: (317) 232-3895
Internet: http://www.ai.org/sema/index.html

Administrator
Iowa Division of Emergency Management
Department of Public Defense
Des Moines, IA 50319
Telephone: (515) 281-3231
Fax: (515) 281-7539
Internet: http://www.state.ia.us/government/dpd/emd/index.htm

Director
Kansas Division of Emergency Preparedness
2800 SW Topeka Boulevard
Topeka, KS 66611-1287
Telephone: (913) 274-1401
Fax: (913) 274-1426
Internet: http://www.ink.org/public/kdem/

Executive Director
State of Kentucky Office of Disaster and Emergency Services
EOC Building, Boone National Guard Center
Frankfort, KY 40601-6168
Telephone: (502) 564-8682
Fax: (502) 564-8614
Internet: http://webserve.dma.state.ky.us

Director
Louisiana Office of Emergency Preparedness
PO Box 44217
Baton Rouge, LA 70804
Telephone: (504) 342-1583
Fax: (504) 342-5471
Internet: http://199.188.3.91/

Director
Maine Emergency Management Agency
State Office Building, Station 72
Augusta, ME 04333
Telephone: (207) 287-4080
Fax: (207) 287-4079
Internet: http://www.state.me.us/mema/memahome.htm

Director
Maryland Emergency Management Agency
Camp Fretterd Military Reservation
5401 Rue Saint Lo Drive
Reistertown, MD 21136
Telephone: (410) 517-3600
Fax: (410) 517-3610
Toll Free: (877) 636-2872
Internet: http://www.mema.state.md.us/

Director
Massachusetts Emergency Management Agency
400 Worcester Road, PO Box 1496
Framingham, MA 01701-0317
Telephone: (508) 820-2010
Fax: (508) 727-4764
Internet: http://www.magnet.state.ma.us/mema/homepage.htm

Director
Michigan Division of Emergency Management
300 South Washington Square, Suite 300
Lansing, MI 48913
Telephone: (517) 334-5103
Fax: (517) 333-4987
Internet: http://www.msp.state.mi.us/division/emd/emdweb1.htm

Director
Minnesota Division of Emergency Management
Department of Public Safety
Suite 2231, 444 Cedar Street
St. Paul, MN 55101-6223
Telephone: (612) 296-0450
Fax: (612) 296-0459
Internet: http://www.dps.state.mn.us/emermgt/

Director
Mississippi Emergency Management Agency
PO Box 4501, Fondren Station
Jackson, MS 39296-4501
Telephone: (601) 352-9100
Fax: (601)352-8314
Internet: http://www.state.ms.us

Director
State of Missouri Emergency Management Agency
PO Box 116, 2302 Militia Drive
Jefferson City, MO 65102
Telephone: (573) 526-9146
Fax: (573) 634-7966
Internet: http://www.sema.state.mo.us/semapage.htm

Administrator
Montana Division of Disaster and Emergency Services
1100 North Main, PO Box 4789
Helena, MT 59604-4789
Telephone: (406) 444-6911
Fax: (406) 444-6965
Internet: http://www.mt.gov/dma/des/index.shtml

Director
Nebraska State Civil Defense Agency
National Guard Center
1300 Military Road
Lincoln, NE 68508-1090
Telephone: (402) 471-7410
Fax: (402) 471-7433
Internet: http://www.nol.org/home/nmd/nema.htm

Chief
Nevada Division of Emergency Management
Capitol Complex, 2525 South Carson Street
Carson City, NV 89701
Telephone: (702) 687-4240
Fax: (702) 687-6788
Internet: http://www.state.nv.us/dmv_ps/emermgt.htm

Director
New Hampshire Governor's Office of Emergency Management
State Office Park South, 107 Pleasant Street
Concord, NH 03301
Telephone: (603) 271-2231
Fax: (603) 225-7341

Deputy State Director
New Jersey Office of Emergency Management
PO Box 7068, Old River Road
West Trenton, NJ 08628-0068
Telephone: (609) 538-6050
Fax: (609) 538-0345
Internet: http://www.state.nj.us/lps/njsp/outfit-p.html

Director
New Mexico Division of Emergency Management
Department of Public Safety
PO Box 1628
Santa Fe, NM 87504-1628
Telephone: (505) 827-9222
Fax: (505)827-3456
Internet: http://www.dps.state.nm.us/

Director
New York State Emergency Management Office
22 Security Building, State Campus
Albany, NY 12226-5000
Telephone: (518) 457-9996
Fax: (518) 457-9995
Internet: http://www.nysemo.state.ny.us/

Director
North Carolina Division of Emergency Management
116 West Jones Street
Raleigh, NC 27603
Telephone: (919) 733-3718
Fax: (919) 733-5406
Internet: http://www.dem.dcc.state.nc.us/

Director
North Dakota Division of Emergency Management
PO Box 5511
Bismarck, ND 58506-5511
Telephone: (701) 328-8100
Fax: (701) 328-8181
Internet: http://www.state.nd.us/dem

Deputy Director
Ohio Emergency Management Agency
2825 W. Dublin Granville Road
Columbus, OH 43235-2206
Telephone: (614) 889-7150
Fax: (614) 889-7183
Internet: http://www.state.oh.us/odps/division/ema/

Director
Oklahoma Department of Civil Emergency Management
PO Box 53365
Oklahoma City, OK 73152
Telephone: (405) 521-2481
Fax: (405) 521-4053
Internet: http://www.onenet.net/~odcem/

Director
Oregon Division of Emergency Management
595 Cottage Street, NE
Salem, OR 97310
Telephone: (503) 378-2911
Fax: (503) 588-1378
Internet: http://www.osp.state.or.us/oem/oem.htm

Director
Pennsylvania Emergency Management Agency
2605 Interstate Drive
Harrisburg, PA 17110-9364
Telephone: (717) 651-2001
Fax: 651-7800
Internet: http://www.pema.state.pa.us/

Director
Rhode Island Emergency Management Agency
645 New London Ave
Cranston, RI 02920-3003
Telephone: 401-946-9996
Fax: (401) 941-1891
Internet: http://www.state.ri.us/riema/riemaaa.html

Director
South Carolina Emergency Preparedness Division
Office of the Adjutant General
1429 Senate Street
Columbia, SC 29201
Telephone: (803) 734-8020
Fax: (803) 734-8062
Internet: http://www.state.sc.us/epd/

Director
South Dakota Division of Emergency Management
500 East Capitol
Pierre, SD 57501-5070
Telephone: (605) 773-3233
Fax: (605) 773-3580
Internet: http://www.state.sd.us/state/executive/military/sddem.htm

Director
Tennessee Emergency Management Agency
3041 Sidco Drive, PO Box 41502
Nashville, TN 37204-1502
Telephone: (615) 741-6528
Fax: (615) 242-9635
Internet: http://www.tnema.org

State Coordinator
Texas Division of Emergency Management
Department of Public Safety
PO Box 4087, North Austin
Austin, TX 78733-0225
Telephone: (512) 465-2443
Fax: (512) 424-2444
Internet: http://www.txdps.state.tx.us/dem/

Director
Utah Division of Comprehensive Emergency Management
State Office Building, Room 1110
Salt Lake City, UT 84114
Telephone: (801) 538-3400
Fax: (801) 538-3770
Internet: http://www.dps.state.ut.us/cem/cem1.htm

Director
Vermont Division of Emergency Management
Waterbury State Complex
103 South Main Street
Waterbury, VT 05671-2101
Telephone: (802) 244-8721
Fax: (802) 244-8655
Internet: http://www.dps.state.vt.us/vem/index.html

Director
Puerto Rico Civil Defense Agency
Office of the Governor
PO Box 5127
San Juan, PR 00906
Telephone: (809) 724-0124
Fax: (809) 725-4244

Deputy Director
Virgin Islands Office of Civil Defense and Emergency Services
102 Estate Atmon
St. Croix, VI 00820
Telephone: (809) 773-2244
Fax: (809) 774-1491

State Coordinator
Virginia Department of Emergency Services
10501 Trade Court
Richmond, VA 23236-3713
Telephone: (804) 897-6502
Fax: (804) 897-6506
Internet: http://www.vdes.state.va.us

Director
State of Washington
Washington Military Department
Emergency Management Division
Building 20, M/S: TA-20
Camp Murray, WA 98430-5122
Telephone: (253) 512-7000
Fax: (253) 512-7200
Internet: http://www.wa.gov/mil/wsem/

State Director
West Virginia Office of Emergency Services
Main Capitol Building, Room EB-80
Charleston, WV 25305-0360
Telephone: (304) 558-5380
Fax: (304) 344-4538
Internet: http://www.state.wv.us/wvoes

Administrator
Wisconsin Division of Emergency Government
2400 Wright Street, PO Box 7865
Madison, WI 53707
Telephone: (608) 242-3232
Fax: (608) 242-3247
Internet: http://badger.state.wi.us/agencies/dma/wem/index.htm

Coordinator
Wyoming Emergency Management Agency
5500 Bishop Boulevard
Cheyenne, WY 82009
Telephone: (307) 777-4920
Fax: (307) 635-6017

Manager
American Samoa Territorial Emergency Management Coordination
Department of Public Safety
PO Box 1086
Pago Pago, American Samoa 96799
Telephone: (011)(684) 633-2331
Fax: (011)(684)633-2300

Director
Guam Division of Civil Defense Emergency Services Office
PO Box 2877
Agana, Guam 96910
Telephone: (011)(671) 477-9841
Fax: (671) 477-3727
Internet: http://ns.gov.gu/

Civil Defense Coordinator
Mariana Islands Office of Civil Defense
Capitol Hill
Saipan, Mariana Islands 96950
Telephone: (011)(670) 322-9529
Fax: (011)(670)322-2545

Civil Defense Coordinator
Republic of the Marshall Islands
PO Box 15
Majuro, Republic of the Marshall Islands 96960
Telephone: (011)(692) 730-3232
Fax: (011)(692)625-3649

Special Assistant to the President of Micronesia Disaster Coordination
Office of the President
PO Box 490
Kolonia, Pohnpei - Micronesia 96941
Telephone: (011)(691) 320-2822
Fax: (011)(691) 320-2785

Palau NEMO Coordinator
Office of the President
PO Box 100
Koror, Republic of Palau 96940
Telephone: (011)(680) 488-2422
Fax: (011)(680) 488-3312

Appendix G

Examples of Tables of Contents for a City Plan for Emergency Response

EXAMPLE 1 – Emergency Response Plan for Terrorist Incidents

Table of Contents:

1. Forward
2. Acknowledgments
3. Concept of Operations
4. Incident Command (direction and control)
5. Hazard Analysis (potential scenarios and effects)
6. Incident Recognition
7. Radiological Terrorism Plan Overview
8. Emergency Management
   Mitigation
   Preparedness
   Response
   Recovery
9. Agency Specific Responsibilities
   Mayor's Office of Emergency Management
   City Fire Department
   City Police Department
   Department of Health
   Department of Corrections
   Hospital/Medical Facilities
   Department of Sanitation
   Public Department of Transportation
   American Red Cross

Federal Agencies:
- Federal Bureau of Investigation
- Federal Emergency Management Agency
- U.S. Department of Defense
- U.S. Department of Health and Human Services
- U.S. Department of Transportation
- Public utilities

10. Appendices
    - Notification Procedures
    - Important Notification Telephone Numbers
    - Federal Assistance
    - National Associations for Emergency Hazardous Materials Assistance
    - Medical Protocol

EXAMPLE 2 – Sample Emergency Operations Plan

Definitions
Base Plan:
- Introduction
  - Purpose
  - Scope
  - Organization
- Policies
  - Authorities
  - Assignment of Responsibilities
  - Resource Coordination
  - Recovery Operations
  - Operations Facilities
  - Public Information
- Situation – Disaster Condition
- Mission
- Concept of Operations
  - General
  - Organization
    - Management Team
    - Emergency Support Team
    - Field Response Team
  - Response Action
  - Responsibilities

Functional Annexes:
- Management and Coordination – City Incident Command System
  - Introduction
  - Situations and Assumptions
  - Mission
  - Execution
  - Concept of Operations
  - Mobilization
  - Individual Duties and Responsibilities
  - Communications
  - Training
- Law Enforcement – Sheriff Department
  - Introduction
  - Situations and Assumptions
  - Mission
  - Execution
  - Concept of Operations
  - Administration
- Fire Fighting/Search and Rescue

Hazardous Materials Response Annex

Health and Medical Annex -
- Patient Tracking
- Medical Decontamination Resources

Communications – Emergency Disaster Notifications

Public Works and Engineering

Information and Planning

Warning

Resources

Recovery

Transportation

Support Annexes:
- Damage Assessment/Damage Declaration
- Public Information/JIC
- Evacuation and Mass Care
- Financial Management
- Transportation

Appendix H

Training Under the Domestic Preparedness Program

The federal government has made available training to help ensure that local and state responders have the knowledge and skills necessary for WMD incidents. The most recent and comprehensive effort in this regard was mandated by Title XIV of Public Law 104-201, the National Defense Authorization Act for Fiscal Year 1997. This legislation, also known as Nunn-Lugar-Domenici, directed the federal government to improve the capabilities of local and state agencies to respond to incidents involving WMD. DOD was directed to lead the Federal Government effort. The U.S. Army Soldier Biological and Chemical Command (SBCCOM) was designated as the Program Director for Domestic Preparedness to coordinate, integrate, and execute a program to enhance domestic preparedness to nuclear, biological or chemical (NBC) terrorism. In 1999, this responsibility was transferred to DOJ. The purpose of this program was to provide for training of local and state emergency responders in the event of a terrorist incident involving NBC WMD. The training program was intended to "train the trainers" and be in the form of modules which could be tailored to meet the specific training needs of individual cities and readily integrated into the existing emergency responder training programs at the local and state level. The following is an outline of the courses that have been developed.

Emergency Responder Awareness – (4 h)

- NBC terrorist threat
- chemical agents
- biological agents

- nuclear materials
- recognizing dissemination devices
- responder action

Emergency Responder Operations – (4 h)

- responder actions
- chemical downwind hazard analysis
- personal protection
- introduction to detection and identification
- emergency decontamination procedures
- operations course exercise

Technician – HAZMAT (12 h)

- NBC terrorist threat
- nuclear materials
- chemical agents
- biological agents
- recognizing dissemination devices
- advanced detection and identification
- detection practical exercise
- personal protection equipment
- decontamination procedures
- simplified downwind hazard prediction
- responder action at the technician level

Technician – Emergency Medical Service (8 h)

- medical aspects of radiological materials
- medical aspects of biological agents
- medical aspects of chemical agents
- special considerations for mass casualty triage and decontamination
- response exercise

Technician – Hospital Provider (8 h)

- introduction to NBC disasters
- medical aspects of radiological agents
- medical aspects of biological agents
- medical aspects of chemical agents

- special considerations for mass casualty triage and decontamination

Incident Command (6 h)

- challenges and consequences of management in a NBC incident
- tactical considerations and actions in an NBC terrorist incident
- understanding the role of the federal government in an NBC terrorist incident
- NBC terrorism response and planning exercise

Senior Leadership (6.5 h)

A workshop intended to instruct and inform the senior leadership of the city.

- employ an integrated planning, training and exercising effort among all involved local agencies and between the local jurisdiction and its mutual aid partners for an NBC terrorist incident
- recognize probable NBC situations and the implications these situations have on the community
- interact with state and federal personnel so the operational assets can be assembled, assigned and employed with maximum effectiveness
- interact with the media to calm public fears and maintain public confidence in local government

The training provided under the Domestic Preparedness program can serve as a baseline which can be built upon for different target audiences. Table H.1 lists currently available federally sponsored courses suitable to address the basic and advanced health physics aspects of crisis and consequence management.

TABLE H.1—*Currently available federally sponsored courses.*

	Course	Sponsor	Agency	Audience	Occupation	Competency Level	Basic/ Refresher/ Advanced
1	NBC DP Training Incident Command	DOD	SBCCOM	C	C	Command	B/R
2	NBC DP Training Responder – Awareness	DOD	SBCCOM	C	F/H/L/E/C	Awareness	B/R
3	NBC DP Training Responder – Operations	DOD	SBCCOM	C	F/H/L/E/C	Operations	B/R
4	NBC DP Training Senior Officials Workshop	DOD	SBCCOM	C	PO/SO	Senior management	B/R
5	NBC DP Training – Emergency Medical Services	DOD	SBCCOM	C	E	Technician/Specialist	B/R
6	NBC DP Training – Technician HAZMAT	DOD	SBCCOM	C	H	Technician/Specialist	B/R
7	NBC DP Training – Hospital Provider	DOD	SBCCOM	C	N(E)/D(E)	Operations/Medical	B/R
8	Medical Effects of Ionizing Radiation	DOD	AFRRI	M	E/N(E,T)/D(E,T)/S	Medical	A
9	Radiological Emergency Team Operations	DOD	DNWS	M	S	Operations/Technician /Specialist	A
10	Radiological Hazards Training Course	DOD	DNWS	M	E/N(E)/D(E)	Technician/Specialist/ Medical	A
11	Radiological Accident Command, Control and Communication	DOD	DNWS	M/DOD C	C	Senior management	A

	Course	Sponsor	Agency	Audience	Occupation	Competency Level	Basic/ Refresher/ Advanced
12	Preparing for and Managing the Consequences of Terrorism	DOD	Guard	B	F/H/C	Awareness/Operations	A
13	Strategic Decision Making and Management of Complex Contingencies	DOD	NDU	B	SO	Senior management	A
14	Hazardous Materials Incident Response Operations	DOE	HR-FORS	B	F/H/L/C/S	Awareness/Operations	B/R/A
15	Handling of Radiation Accidents by Emergency Personnel	DOE	REAC/TS	B	F/H/E	Technician/Specialist	A
16	Health Physics in Radiation Accidents	DOE	REAC/TS	B	E/N(E)/D(E)/S	Technician/Specialist	A
17	Medical Planning and Care in Radiation Accidents	DOE	REAC/TS	C	E/N(E,T)/D(E,T)	Operations/Medical	A
18	Radioactive Material Basics for Emergency Responders	DOE	EM-76	B	F/H/L/C	Awareness/Operations /Senior management	A
19	Radiological Emergency Response	DOE	ORISE	B	F/H/E/L	Awareness/Operations /Technician	A
20	Nuclear Emergency Planning	DHHS	HSPH	B	C/PO/SO/S	Senior management	A
21	Advanced Radiation Incident Operations	FEMA	EMI	B	S	Operations	A
22	Integrated Emergency Management: Consequences of Terrorism	FEMA	EMI	B	C/PO/SO/S	Operations	A

23	Incident Command System/Emergency Operations Center Interface	FEMA	EMI	C	C/PC/SO	Operations	A
24	Emergency Response to Criminal/Terrorist Incidents	FEMA	EMI	C	F/H/L/E	Operations	A
25	Fundamentals Course for Radiological Response Teams	FEMA	EMI	C	F/H/S	Technician/Specialist	A
26	Fundamentals Course for Radiological Monitors	FEMA	EMI	C	F/H/S	Technician/Specialist	A
27	Mass Fatalities Incident Course	FEMA	EMI	B	F/H/C/L	Operations	A
28	Radiological Emergency Response Operations	FEMA	EMI	B	F/H/E/S	Operations	A
29	Emergency Response to Terrorism: Basic Concepts	FEMA	NFA	B	F/H/L/E	Operations	A
30	Emergency Response to Terrorism: Incident Management	FEMA	NFA	B	C	Senior management	A
31	Emergency Response to Terrorism: Self-Study	FEMA	NFA	B	F/H/L	Awareness	A
32	Emergency Response to Terrorism: Tactical Considerations	FEMA	NFA	C	F/H/E/N(E)/D(E)/C	Technician/Specialist	A
33	Incident Command System for Emergency Medical Services	FEMA	NFA	C	E	Technician/Specialist	A
34	First Responder Training Workshop: Public Transportation Chemical, Biological and Nuclear Incidents	DOT	R/SPA	C	T	Awareness/Specialist	A

Sponsor
- DHHS – U.S. Department of Health and Human Services
- DOD – U.S. Department of Defence
- DOE – U.S. Department of Energy
- DOT – U.S. Department of Transportation
- FBI – Federal Bureau of Investigation
- FEMA – Federal Emergency Management Agency

Agency
- AFRRI – Armed Forces Radiobiology Research Institute
- DNWS – Defense Nuclear Weapon School
- EMI – Emergency Management Institute
- Guard – National Guard Bureau
- HSPH – Harvard School of Public Health
- NDU – National Defense University
- NFA – National Fire Academy
- ORISE – Oak Ridge Institute for Science and Education
- REAC/TS – Radiation Emergency Assistance Center/Training Site
- R/SPA – Research/Special Program Administration
- SBCCOM – U.S. Army Soldier Biological and Chemical Command

Audience
- C – Civilian
- M – Military
- B – Both

Occupation
- C – Incident commander
- D – Doctor; (E) – Emergency, (T) – Tertiary
- E – EMT/Paramedic
- F – Firefighters
- H – HAZMAT
- L – Law enforcement
- N – Nurse; (E) – Emergency, (T) – Tertiary
- P – Police
- PO – Public official
- S – Special teams
- SO – Senior official
- T – Transportation

Glossary

ALARA (as low as reasonably achievable): A principle of radiation protection philosophy that requires that exposures to ionizing radiation should be kept as low as reasonably achievable, economic and social factors being taken into account. The ALARA principle is equivalent to the principle of optimization defined by the ICRP which states that protection from radiation exposure is optimum when the expenditure of further resources would be unwarranted by the reduction in exposure that would be achieved.

ambient dose equivalent: The dose equivalent at a point in a radiation field that would be produced by the corresponding expanded and aligned field in the ICRU sphere at a depth (*d*) on the radius opposing the direction of the aligned field. See ICRU Report 51 (ICRU, 1993) for a more detailed discussion of this quantity. The special name for the unit of ambient dose equivalent is sievert (Sv).

absorbed dose (*D*)**:** The quotient of $d\bar{\varepsilon}$ by dm, where $d\bar{\varepsilon}$ is the mean energy imparted by ionizing radiation to the matter in a volume element and dm is the mass of the matter in that volume element, *i.e.*, the absorbed dose, $D = d\bar{\varepsilon}/dm$. The unit of absorbed dose is the gray (Gy).

acute radiation syndrome: A broad term used to describe a range of signs and symptoms that reflect severe damage to specific organ systems that can lead to death within hours or several weeks.

consequence management: A public health and safety function that includes measures taken to minimize the impact of a disaster on the public and the environment.

crisis management: A law enforcement function that includes measures to identify, acquire, and plan the use of resources to anticipate, prevent, and/or resolve a threat or act of terrorism.

deterministic effects: Effects for which the severity of the effect in affected individuals varies with the dose, and for which a threshold usually exists.

dose equivalent (H): The product of quality factor (*Q*) and *D* at a point in tissue, where *D* is the absorbed dose and *Q* is the dimensionless quality factor at that point, thus

$H = Q\,D$.(G.2)

The unit of dose equivalent is joule per kilogram ($J\ kg^{-1}$) and its special name is the sievert (Sv).

effective dose (E): The sum over specified tissues of the products of the equivalent dose in a tissue (T) and the weighting factor for that tissue

(w_T). That is, where E is the effective dose, H_T is the equivalent dose to tissue (T), and w_T expresses the fraction of the total stochastic risk associated with the irradiation of tissue (T). The special name for the unit of effective dose is the sievert (Sv); formerly the unit was the rem (1 Sv = 100 rem).

equivalent dose (H_T): A quantity used for radiation protection purposes that takes into account the different probability of effects which occur with the same absorbed dose delivered by radiations with different radiation weighting factors. It is defined as the product of the average absorbed dose in a specified organ or tissue (D_T) and the radiation weighting factor (w_R). The unit of equivalent dose is joule per kilogram (J kg^{-1}) and its special name is the sievert (Sv).

external exposure: An exposure received from a source of ionizing radiation outside of the body.

improvised nuclear device: A device designed by terrorists to produce a nuclear detonation. This includes stolen and subsequently modified nuclear weapons but does not include stockpiled weapons in the custody of the military.

internal exposure: An exposure received from a source of ionizing radiation inside of the body.

Joint Information Center (JIC): The primary location for the coordination and release of federal, state and local government and voluntary and private responding organization media relations, community relations, and internal information.

latent period: The period of time between exposure to ionizing radiation and the appearance of radiation effects.

quality factor (Q): A multiplying factor used with absorbed dose to convert to dose equivalent and therefore to express the radiation's effectiveness in causing stochastic effects.

radiation weighting factor: A factor used for radiation-protection purposes that accounts for differences in biological effectiveness between different radiations. The radiation weighting factor (w_R) is independent of the tissue weighting factor (w_T).

radiological dispersal device (RDD): A device designed to spread radioactive material through a detonation of conventional explosives or other (non-nuclear) means.

spent fuel: Fuel that has been withdrawn from a nuclear reactor following irradiation, the constituent elements of which have not been separated by reprocessing.

stochastic effects: Effects, the probability of which, rather than their severity is a function of dose without a threshold.

tissue weighting factor (w_T): A factor for a particular tissue representing the fraction of the detriment (cancer) plus hereditary effects) attributed to that tissue when the whole body is irradiated uniformly.

Acronyms

AFRRI	Armed Forces Radiobiology Research Institute
ALARA	As low as reasonably achievable
APA	American Psychiatric Association
CMRT	Consequence Management Response Team
CNS	Central nervous system
CPX	Command post exercise
DHHS	U.S. Department of Health and Human Services
DNA	Deoxyribonucleic acid
DOD	U.S. Department of Defense
DOE	U.S. Department of Energy
DOJ	U.S. Department of Justice
DOS	U.S. Department of State
DOT	U.S. Department of Transportation
EPA	U.S. Environmental Protection Agency
ESF	Emergency Support Functions
FBI	Federal Bureau of Investigation
FCE	Forward Command Element
FDA	Food and Drug Administration
FEMA	Federal Emergency Management Agency
FEST	Foreign Emergency Support Team
FRERP	Federal Radiological Emergency Response Plan
FRMAC	Federal Radiological Monitoring and Assessment Center
FRP	Federal Response Plan
FTX	Field training exercises
GAO	U.S. Government Accounting Office
HAZMAT	Hazardous material
IAEA	International Atomic Energy Agency
IC	Incident commander
ICP	Incident command post
ICS	Incident Command System
ICRP	International Commission on Radiological Protection
ICRU	International Commission on Radiation Units and Measurements
IOM/NRC	Institute of Medicine/National Research Council
IQ	Intelligence quotient
JIC	Joint Information Center
JOC	Joint Operations Center
KI	Potassium iodide

LFA	Lead federal agency
MAPEX	Map, or tabletop, exercises
MRAT	Medical Radiobiology Advisory Team
NAS/NRC	National Academy of Sciences/National Research Council
NBC	Nuclear, biological or chemical
NCP	National Oil and Hazardous Substances Pollution Contingency Plan
NCRP	National Council on Radiation Protection and Measurements
NRC	Nuclear Regulatory Commission
OSC	On-Scene Commander
PDD	Presidential Decision Directive
PHS	U.S. Public Health Service
PTSD	Post-Traumatic Stress Disorder
RDD	Radiological dispersal device
REAC/TS	Radiation Emergency Assistance Center/Training Site
SBCCOM	U.S. Army Soldier Biological and Chemical Command
STX	Situational training exercises
TMI	Three Mile Island
UNSCEAR	United Nations Scientific Committee on the Effects of Atomic Radiation
WMD	Weapon of mass destruction

Conversions of Conventional and International System of Dosimetric Quantities

In the United States absorbed dose is commonly given in rad and other protection quantities, such as equivalent dose and effective dose, are given in rem. However, the NCRP (1985) supports the use of the International System [Le Système International Unités (SI)] of Units, in which absorbed dose is given in gray (Gy) and other protection quantities are given in sievert (Sv). The following table is provided to help avoid confusion among persons not familiar with these quantities. The use of the newer system of units would be particularly useful during radiological incidents involving international responders.

Conversions for absorbed dose:

0.001 rad	= 1 mrad	= 0.01 mGy	
0.01 rad	= 10 mrad	= 0.1 mGy	
0.1 rad	= 100 mrad	= 1 mGy	= 0.001 Gy
1 rad	= 1,000 mrad	= 10 mGy	= 0.01 Gy
10 rad		= 100 mGy	= 0.1 Gy
100 rad		= 1,000 mGy	= 1 Gy
1,000 rad			= 10 Gy

Conversions for effective dose, equivalent dose, dose equivalent, and ambient dose equivalent:

0.001 rem	= 1 mrem	= 0.01 mSv	
0.01 rem	= 10 mrem	= 0.1 mSv	
0.1 rem	= 100 mrem	= 1 mSv	= 0.001 Sv
1 rem	= 1,000 mrem	= 10 mSv	= 0.01 Sv
10 rem		= 100 mSv	= 0.1 Sv
100 rem		= 1,000 mSv	= 1 Sv
1,000 rem			= 10 Sv

Many radiation survey instruments in the United States are currently calibrated for use in gamma- or x-ray fields using the exposure quantity roentgen (R). For these types of radiation environments, it is acceptable to temporarily interpret a reading on a survey instrument of 1 R as an absorbed dose of 0.01 Gy, or as an ambient dose equivalent of 0.01 Sv. Similarly, for a pocket dosimeter or other personal dosimetry devices worn on the body and calibrated in roentgens, a reading of 1 R may be interpreted as an absorbed dose of 0.01 Gy or a dose equivalent of 0.01 Sv. It is also acceptable in these environments to temporarily assume that the numerical value of an ambient dose equivalent measured by a survey instrument in a gamma- or x-ray field is equivalent to a measurement of the effective dose that would be received by an individual at that location. The final assessment of the effective dose received by an individual should be made by taking into account all relevant knowledge of the radiation environment and the response characteristics of available dosimeters.

References

AFRRI (1999). Armed Forces Radiobiology Research Laboratory. *Medical Management of Radiological Casualties Handbook*, AFRRI Special Publication 99-2 (Armed Forces Radiobiology Research Laboratory, Bethesda, Maryland).

APA (1994). American Psychiatric Association. *DSM-IV Casebook: A Learning Companion to the Diagnostic and Statistical Manual of Mental Disorders*, 4th ed. (American Psychiatric Press, Washington).

BAUM, A. and FLEMING, I. (1993). "Implications of psychological research on stress and technological accidents," Am. Psychol. **48**, 665-672.

BAUM, A., FLEMING, R. and DAVIDSON, L.M. (1983). "Natural disaster and technological catastrophe," Environ. Behavior **15**, 333-354.

BECKER, S.M. (1997). "Psychosocial assistance after environmental accidents: A policy perspective," Environ. Health Perspect. **105**, 1557–1563.

BECKER, S.M. (2000). "Environmental disaster education at the university level: An integrative approach," Safety Science **35**, 95–104.

BECKER, S.M. (2001a). "Addressing the psychosocial consequences of nuclear disasters: Program and policy issues," pages 9 to 18 in *Medical Report: International Medical Collaboration for Nuclear Disaster*, Haraguchi, Y., Tomoyasu, Y., Nagata, T. and Arai, T., Eds. (National Hospital Tokyo Disaster Medical Center, Tokyo).

BECKER, S.M. (2001b). "Psychosocial effects of radiation accidents," pages 517 to 523 in *Medical Management of Radiation Accidents*, 2nd ed., Gusev, I., Guskova, A.K. and Mettler, F.A., Jr., Eds. (CRC Press, Boca Raton, Florida).

BENNETT, P. (1999). "Understanding responses to risk: Some basic findings," pages 3 to 19 in *Risk Communication and Public Health*, Bennett, P. and Calman, K., Eds. (Oxford University Press, New York).

BOWLER, R.M., MERGLER, D., HUEL, G. and CONE, J.E. (1994). "Psychological, psychosocial, and psychophysiological sequelae in a community affected by a railroad chemical disaster," J. Trauma. Stress **7**, 601-624.

BROMET, E.J., PARKINSON, D.K. and DUNN, L.O. (1990). "Long-term mental health consequences of the accident at Three Mile Island," Int. J. Ment. Health **19**, 48-60.

BROWNE, D., WEISS, J.F., MACVITTIE, T.J. and PILLAI, M.V., Eds. (1989). *Treatment of Radiation Injuries* (Plenum Press, New York).

CAPIZZI, R.L. (1999). "Clinical status and optimal use of amifostine," Oncology **13**, 47-59.

CARTER, B.J. and CAMMERMEYER, M. (1985). "Emergence of real casualties during simulated chemical warfare training under high heat conditions," Mil. Med. **150**, 657-665.

COLLINS, D.L. (1991). "Stress at Three Mile Island: Altered perceptions, behaviors and neuroendocrine values," pages 71 to 79 in *The Medical Basis for Radiation-Accident Preparedness III: Psychological Perspectives*, Ricks, R.C. and Berger, M.E., Eds. (Elsevier Science, New York).

COLLINS, D.L. and DE CARVALHO, A.B. (1993). "Chronic stress from the Goiania ^{137}Cs radiation accident," Behav. Med. **18**, 149-157.

CONKLIN, J.J. and WALKER, R.I., Eds. (1987). *Military Radiobiology* (Academic Press, Inc., New York).

COVELLO, V.T., MCCALLUM, D.B. and PAVLOVA, M.T., Eds. (1989). *Effective Risk Communication: The Role and Responsibility of Government and Nongovernment Organizations* (Plenum Press, New York).

DE CARVALHO, A.B. (1999). "Psychological aspects of a radiological accident," paper presented at the Ninth Annual National Radiological Preparedness Conference, March 29-31, 1999, Baton Rouge, Louisiana.

DHHS (1996). U.S. Department of Health and Human Services. *Responding to the Needs of People with Serious and Persistent Mental Illness in Times of Major Disaster*, DHHS Publication No. (SMA) 96-3077 (National Mental Health Services Knowledge Exchange Network, Washington).

DIGIOVANNI, C., JR. (1999). "Domestic terrorism with chemical or biological agents: Psychiatric aspects," Am. J. Psychiatry **156**, 1500–1505.

DOLL, R. and WAKEFORD, R. (1997). "Risk of childhood cancer from fetal irradiation," Br. J. Radiol. **70**, 130-139.

EDELSTEIN, M.R. (1988). *Contaminated Communities: The Social and Psychological Impacts of Residential Toxic Exposure* (Westview Press, Boulder, Colorado).

EPA (1992). U.S. Environmental Protection Agency. *Manual of Protective Action Guides and Protective Actions for Nuclear Incidents*, EPA 400-R-92-001 (U.S. Environmental Protection Agency, Washington).

EPA (1994). U.S. Environmental Protection Agency. *National Oil and Hazardous Substances Pollution Contingency Plan*, Code of Federal Regulations 40CFR300 [document can be obtained electronically at http:\\frwebgate.access.gpo.gov] (U.S. Government Printing Office, Washington).

ERIKSON, K. (1994). *A New Species of Trouble: The Human Experience of Modern Disasters* (W.W. Norton and Company, New York).

EVERLY, G.S., JR. and MITCHELL, J.T. (1999). *Critical Incident Stress Management–CISM–: A New Era and Standard of Care in Crisis Intervention*, 2nd ed. (Chevron Publishing Corporation, Ellicott City, Maryland).

FARBEROW, N.L. and GORDON, N.S. (1981). *Manual for Child Health Workers in Major Disasters*, DHHS Publication No. (ADM) 86-1070 (National Mental Health Services Knowledge Exchange Network, Washington).

FBI (1998). Federal Bureau of Investigation. *Weapons of Mass Destruction Incident Contingency Plan* (Federal Bureau of Investigation, Washington).

FDA (1998). Food and Drug Administration. "Accidental radioactive contamination of human and animal feeds: Recommendations for state and local agencies, notice of availability," pages 43402 to 43403 in *Federal Register* **63**, No. 156 [document can be obtained electronically at http://www.fda.gov/cdrh] (Center for Devices and Radiological Health, Rockville, Maryland).

FEMA (1996). Federal Emergency Management Agency. "Federal Radiological Emergency Response Plan (FRERP); operational plan; notice," pages 20944 to 20970 in *Federal Register* **61**, No. 90, Part III [document can be obtained electronically at http:\\frwebgate.access.gpo.gov] (U.S. Government Printing Office, Washington).

FEMA (1998). Federal Emergency Management Agency. *FEMA Emergency Information Field Guide* (Federal Emergency Management Agency, Washington).

FEMA (1999). Federal Emergency Management Agency. *The Federal Response Plan*, FEMA-9230 (National Technical Information Service, Springfield, Virginia).

FIGLEY, C.R. (1986). "Traumatic stress: The role of the family and social support system," pages 39 to 56 in *Trauma and Its Wake: Volume II: Traumatic Stress Theory, Research, and Intervention*, Brunner/Mazel Psychosocial Stress Series No. 8, Figley, C.R., Ed. (Brunner/Mazel Publishers, New York).

FLYNN, B.W. (1996). "Psychological aspects of terrorism," presented at the First Harvard Symposium on the Medical Consequences of Terrorism, April 24-25, 1996, Boston, Massachusetts.

FULLERTON, C.S. and URSANO, R.J. (1994). "Health care delivery in the high-stress environment of chemical and biological warfare," Mil. Med. **159**, 524-528.

FULLERTON, C.S. and URSANO, R.J. (1997a). "Posttraumatic responses in spouse/significant others of disaster workers," pages 59 to 75 in *Posttraumatic Stress Disorder: Acute and Long-Term Responses to Trauma and Disaster*, Fullerton, C.S. and Ursano, R.J., Eds. (American Psychiatric Press, Inc., Washington).

FULLERTON, C.S. and URSANO, R.J. (1997b). "The other side of chaos: Understanding the patterns of posttraumatic responses," pages 3 to 18 in *Posttraumatic Stress Disorder: Acute and Long-Term Responses to Trauma and Disaster*, Fullerton, C.S. and Ursano, R.J., Eds. (American Psychiatric Press, Inc., Washington).

GAO (1997). U.S. General Accounting Office. *Report to Congressional Requesters. Combating Terrorism: Federal Agencies' Efforts to*

Implement National Policy and Strategy, GAO/NSIAD-97-254 (U.S. General Accounting Office, Washington).

GLASSTONE, S. and DOLAN, P.J., Eds. (1977). *The Effects of Nuclear Weapons*, 3rd ed. (U.S. Department of Defense and U.S. Department of Energy, Washington).

GOANS, R.E., HOLLOWAY, E.C., BERGER, M.E. and RICKS, R.C. (1997). "Early dose assessment following severe radiation accidents," Health Phys. **72**, 513-518.

GOTTLOBER, P., KRAHN, G., KORTING, H.C., STOCK, W. and PETER, R.U. (1996). "The treatment of cutaneous radiation-induced fibrosis with pentoxifylline and vitamin E. An empirical report," Strahlenther Onkol **172**, 34-38.

GRDINA, D.J., SHIGEMATSU, N., DALE, P., NEWTON, G.L., AGUILERA, J.A. and FAHEY, R.C. (1995). "Thiol and disulfide metabolites of the radiation protector and potential chemopreventive agent WR-2721 are linked to both its anti-cytotoxic and anti-mutagenic mechanisms of action," Carcinogenesis **16**, 767-774.

HARTSOUGH, D.M. and MYERS, D.G. (1985). *Disaster Work and Mental Health: Prevention and Control of Stress Among Workers*, DHHS Publication No. (ADM) 87-1422 (National Mental Health Services Knowledge Exchange Network, Washington).

HAVENAAR, J.M., VAN DEN BRINK, W., VAN DEN BOUT, J., KASYANENKO, A.P., POELIJOE, N.W., WOHLFARTH, T. and MEIJLER-ILJINA, L.I. (1996). "Mental health problems in the Gomel region (Belarus): An analysis of risk factors in an area affected by the Chernobyl disaster," Psychol. Med. **26**, 845-855.

HAVENAAR, J.M., RUMYANTZEVA, G.M., VAN DEN BRINK, W., POELIJOE, N.W., VAN DEN BOUT, J., VAN ENGELAND, H. and KOETER, M.W.J. (1997). "Long-term mental health effects of the Chernobyl disaster: An epidemiologic survey in two former Soviet regions," Am. J. Psychiatry **154**, 1605-1607.

HEDDINGS, R.E. (1999). *U.S. Roles in Providing Humanitarian Assistance Following NBD Accidents/Incidents: The Legal Considerations* (Institute for National Security Studies, U.S. Air Force Academy, Colorado Springs, Colorado).

HODGKINSON, P.E. and STEWART, M. (1998). *Coping with Catastrophe: A Handbook of Post-Disaster Psychosocial Aftercare*, 2nd ed. (Routledge, New York).

HOMANN, S.G. and WILSON, D.V. (1995). *HOTSPOT Training Manual: Health Physics Codes for the PC*, UCRL-MA-118617 (Lawrence Livermore National Laboratory, Livermore, California).

IAEA (1980). International Atomic Energy Agency. *Basic Requirements for Personnel Monitoring*, Safety Series No. 14 (International Atomic Energy Agency, Vienna).

IAEA (1987). International Atomic Energy Agency. *Convention on Assistance in the Case of a Nuclear Accident or Radiological Emergency,*

25 I.L.M. 1377, Volume 25, Number 6 (Part 3 of 13) (International Atomic Energy Agency, Vienna).

IAEA (1988a). International Atomic Energy Agency. *The Radiological Accident in Goiania*, STI/PUB/815 (International Atomic Energy Agency, Vienna).

IAEA (1988b). International Atomic Energy Agency. *Emergency Response Planning and Preparedness for Transport Accidents Involving Radioactive Materials*, Safety Series No. 87 (International Atomic Energy Agency, Vienna).

IAEA (1989a). International Atomic Energy Agency. *Recommendations, Emergency Planning and Preparedness for Accidents Involving Radioactive Materials Used in Medicine, Industry, Research and Teaching*, Safety Series No. 91 (International Atomic Energy Agency, Vienna).

IAEA (1989b). International Atomic Energy Agency. *Principles and Techniques for Post-Accident Assessment and Recovery in a Contaminated Environment of a Nuclear Facility*, Safety Series No. 97 (International Atomic Energy Agency, Vienna).

IAEA (1994). International Atomic Energy Agency. *Intervention Criteria in a Nuclear or Radiation Emergency*, Safety Series No. 109 (International Atomic Energy Agency, Vienna).

IAEA (1996). International Atomic Energy Agency. *International Basic Safety Standards for Protection Against Ionizing Radiation and for the Safety of Radiation Sources*, Safety Series No. 115 (International Atomic Energy Agency, Vienna).

ICRP (1977). International Commission on Radiological Protection. *Recommendations of the International Commission on Radiological Protection*, ICRP Publication 26, Annals of the ICRP **1** (Elsevier Science, New York).

ICRP (1984). International Commission on Radiological Protection. *Protection of the Public in the Event of Major Radiation Accidents: Principles for Planning*, ICRP Publication 40, Annals of the ICRP **14** (Elsevier Science, New York).

ICRP (1991). International Commission on Radiological Protection. *1990 Recommendations of the International Commission of Radiological Protection*, ICRP Publication 60, Annals of the ICRP **21** (Elsevier Science, New York).

ICRP (1993). International Commission on Radiological Protection. *Principles for Intervention for Protection of the Public in a Radiological Emergency*, ICRP Publication 63, Annals of the ICRP **22** (Elsevier Science, New York).

ICRP (2000). International Commission on Radiological Protection. *Pregnancy and Medical Radiation*, ICRP Publication 84, Annals of the ICRP **30** (Elsevier Science, New York).

ICRU (1993). International Commission on Radiological Units and Measurements. *Quantities and Units in Radiation Protection Dosimetry*, ICRU Report 51 (International Commission on Radiological Units and Measurements, Bethesda, Maryland).

IOM/NRC (1999a). Institute of Medicine/National Research Council. *Potential Radiation Exposure in Military Operations: Protecting the Soldier Before, During and After*, Committee on Battlefield Radiation Exposure Criteria (National Academy Press, Washington).

IOM/NRC (1999b). Institute of Medicine/National Research Council. *Chemical and Biological Terrorism: Research and Development to Improve Civilian Medical Response* (National Academy Press, Washington).

KASPERSON, R.E. and KASPERSON, J.X. (1996). "The social amplification and attenuation of risk," Annals AAPSS **545**, 95-105.

KNUDSEN, L.B. (1991). "Legally induced abortions in Denmark after Chernobyl," Biomed. Pharmacother **45**, 229-231.

KOSCHEYEV, V.S., LEON, G.R. and GREAVES, I.A. (1997). "Lessons learned and unsolved public health problems after large-scale disasters," Prehospital Disaster Med. **12**, 120-131.

KROLL-SMITH, J.S. and COUCH, S.R. (1993). "Technological hazards: social responses as traumatic stressors," pages 79 to 91 in *The International Handbook of Traumatic Stress Syndromes*, Wilson, J.P. and Raphael, B., Eds. (Plenum Press, New York).

LEONE, B., Ed. (1996). *Urban Terrorism* (Greenhaven Press Inc., San Diego, California).

LEVIN, S.G., YOUNG, R.W. and STOHLER, R.L. (1992). "Estimation of median human lethal radiation dose computed from data on occupants of reinforced concrete structures in Nagasaki, Japan," Health Phys. **63**, 522-531.

LIFTON, R.J. (1967). *Death in Life: Survivors of Hiroshima* (Basic Books, Inc., New York).

LILLIBRIDGE, S.R. and SIDELL, F.R. (1995). *A Report on the Casualties from the Tokyo Subway Incident by the U.S. Medical Team* (Centers for Disease Control and Prevention, Atlanta, Georgia).

LONG, D.E. (1990). *The Anatomy of Terrorism* (The Free Press, New York).

MACVITTIE, T.J., WEISS, J.F. and BROWNE, D., Eds. (1996). *Advances in the Biosciences, Volume 94. Advances in the Treatment of Radiation Injuries* (Elsevier Science, New York).

METTLER, F.A., JR. and UPTON, A.C. (1995). *Medical Effects of Ionizing Radiation*, 2nd ed. (W. B. Saunders Company, Philadelphia).

MICKLEY, G.A. (1989). "Psychological factors in nuclear warfare," page 153 in *Medical Consequences of Nuclear Warfare: Textbook of Military Medicine Series, Part I, Warfare, Weaponry, and the Casualty*, Walker, R.I. and Cerveny, T.J., Eds. (TMM Publications, Office of the Surgeon General, Walter Reed Army Medical Center, Washington).

MILLER, A.H. (1982). *Terrorism, The Media and the Law* (Transnational Publishers, Inc., Dobbs Ferry, New York).

MITCHELL, J.T. and EVERLY, G.S., JR. (1996). *Critical Incident Stress Debriefing: CISD. An Operations Manual for the Prevention of*

Traumatic Stress Among Emergency Service and Disaster Workers, 2nd ed. (Chevron Publishing Corporation, Ellicott City, Maryland).

MYERS, D. (1994). *Disaster Response and Recovery: A Handbook for Mental Health Professionals* (Center for Mental Health Services, Rockville, Maryland).

NAS/NRC (1988). National Academy of Sciences/National Research Council. *Health Risks of Radon and Other Internally Deposited Alpha-Emitters*, Committee on the Biological Effects of Ionizing Radiation, BEIR-IV (National Academy Press, Washington).

NAS/NRC (1990). National Academy of Sciences/National Research Council. *Health Effects of Exposure to Low Levels of Ionizing Radiation*, Committee on the Biological Effects of Ionizing Radiation, BEIR V (National Academy Press Washington).

NCRP (1962). National Council on Radiation Protection and Measurements. *Exposure to Radiation in an Emergency*, NCRP Report No. 29 (superseded by NCRP Report No. 42) (out of print).

NCRP (1974). National Council on Radiation Protection and Measurements. *Radiological Factors Affecting Decision-Making in a Nuclear Attack*, NCRP Report No. 42 (National Council on Radiation Protection and Measurements, Bethesda, Maryland).

NCRP (1977). National Council on Radiation Protection and Measurements. *Protection of the Thyroid Gland in the Event of Releases of Radioiodine*, NCRP Report No. 55 (National Council on Radiation Protection and Measurements, Bethesda, Maryland).

NCRP (1980a). National Council on Radiation Protection and Measurements. *Krypton-85 in the Atmosphere-With Specific Reference to the Public Health Significance of the Proposed Controlled Release at Three Mile Island*, NCRP Commentary No. 1 (National Council on Radiation Protection and Measurements, Bethesda, Maryland).

NCRP (1980b). National Council on Radiation Protection and Measurements. *Management of Persons Accidentally Contaminated with Radionuclides*, NCRP Report No. 65 (National Council on Radiation Protection and Measurements, Bethesda, Maryland).

NCRP (1982). National Council on Radiation Protection and Measurements. *The Control of Exposure of the Public to Ionizing Radiation in the Event of Accident or Attack*, NCRP Symposium Proceedings No. 1 (National Council on Radiation Protection and Measurements, Bethesda, Maryland).

NCRP (1985). National Council on Radiation Protection and Measurements. *SI Units in Radiation Protection and Measurements*, NCRP Report No. 82 (National Council on Radiation Protection and Measurements, Bethesda, Maryland).

NCRP (1987). National Council on Radiation Protection and Measurements. *Ionizing Radiation Exposure of the Population of the United States*, NCRP Report No. 93 (National Council on Radiation Protection and Measurements, Bethesda, Maryland).

NCRP (1993a). National Council on Radiation Protection and Measurements. *Limitation of Exposure to Ionizing Radiation*, NCRP Report No. 116 (National Council on Radiation Protection and Measurements, Bethesda, Maryland).

NCRP (1993b). National Council on Radiation Protection and Measurements. *Risk Estimates for Radiation Protection*, NCRP Report No. 115 (National Council on Radiation Protection and Measurements, Bethesda, Maryland).

NCRP (1994). National Council on Radiation Protection and Measurements. *Advising the Public about Radiation Emergencies: A Document for Public Comment*, NCRP Commentary No. 10 (National Council on Radiation Protection and Measurements, Bethesda, Maryland).

NCRP (1997). National Council on Radiation Protection and Measurements. *Uncertainties in Fatal Cancer Risk Estimates Used in Radiation Protection*, NCRP Report No. 126 (National Council on Radiation Protection and Measurements, Bethesda, Maryland).

NCRP (1998). National Council on Radiation Protection and Measurements. *Radionuclide Exposure of the Embryo/Fetus*, NCRP Report No. 128 (National Council on Radiation Protection and Measurements, Bethesda, Maryland).

NCRP (1999). National Council on Radiation Protection and Measurements. *Recommended Screening Limits for Contaminated Surface Soil and Review of Factors Relevant to Site-Specific Studies*, NCRP Report No. 129 (National Council on Radiation Protection and Measurements, Bethesda, Maryland).

NORTHRUP, J. (1996). *Handbook of Nuclear Weapon Effects* (Defense Special Weapons Agency, Alexandria, Virginia).

NOTEBOOM, J.L., HUMMEL, W.A., BROERSE, J.J., DE VIJLDER, J.J., VULSMA, T., JANSEN, J.T. and VAN BEKKUM, D.W. (1997). "Protection of the maternal and fetal thyroid from radioactive contamination by the administration of stable iodide during pregnancy. An experimental evaluation in chimpanzees," Radiat. Res. **147**, 691-697.

NRC (1980). U.S. Nuclear Regulatory Commission and Federal Emergency Management Agency (Joint Publication). *Criteria for Preparation and Evaluation of Radiological Emergency Response Plans and Preparedness in Support of Nuclear Power Plants*, NUREG-0654, FEMA-REP-1. (U.S. Nuclear Regulatory Commission, Washington).

NRC (2000). U.S. Nuclear Regulatory Commission, U.S. Department of Energy, U.S. Environmental Protection Agency, and U.S. Department of Defense. *Multi-Agency Radiation Survey and Site Investigation Manual (MARSSIM)*, Rev. 1, NUREG-1575, Rev. 1; EPA 402-R-97-016, Rev. 1; DOE/EH-0624, Rev. 1 (National Technical Information Service, Springfield, Virginia).

O'BRIEN, L.S. (1998). *Traumatic Events and Mental Health* (Cambridge University Press, Cambridge, New York).

PDD (1995/1998). Presidential Decision Directive. *U.S. Policy on Counterterrorism*, PDD-39; reaffirmed in *Protection Against Unconventional*

Threats to the Homeland and Americans Overseas, PDD-62 (National Security Council, Washington).

PETER, R.U., GOTTLOBER, P., NADESHINA, N., KRAHN, G., BRAUN-FALCO, O. and PLEWIG, G. (1999). "Interferon gamma in survivors of the Chernobyl power plant accident: New therapeutic option for radiation-induced fibrosis," Int. J. Radiat. Oncol. Biol. Phys. **45**, 147-152.

PIERCE, D.A. and PRESTON, D.L. (2000). "Radiation-related cancer risks at low doses among atomic bomb survivors," Radiat. Res. **154**, 178-186.

PIERCE, D.A., SHIMIZU, Y., PRESTON, D.L., VAETH, M. and MABUCHI, K. (1996). "Studies of the mortality of atomic bomb survivors. Report 12, Part I. Cancer: 1950-1990," Radiat. Res. **146**, 1-27.

PYNOOS, R.S., GOENJIAN, A.K. and STEINBERG, A.M. (1998). "A public mental health approach to the postdisaster treatment of children and adolescents," Child Adolesc. Psychiatr. Clin. N. Am. **7**, 195–210.

RICKS, R.C. and BERGER, M.E., Eds. (1991). *The Medical Basis for Radiation-Accident Preparedness III: Psychological Perspectives* (Elsevier Science, New York).

RON, E., LUBIN, J.H., SHORE, R.E., MABUCHI, K., MODAN, B., POTTERN, L.M., SCHNEIDER, A.B., TUCKER, M.A. and BOICE, J.D., JR. (1995). "Thyroid cancer after exposure to external radiation: A pooled analysis of seven studies," Radiat. Res. **141**, 259-277.

ROSA, E.A. and FREUDENBURG, W.R. (1993). "The historical development of public reactions to nuclear power: Implications for nuclear waste policy," pages 32 to 63 in *Public Reactions to Nuclear Waste: Citizens' Views of Repository Siting*, Dunlap, R.E., Kraft, M.E. and Rosa, E.A., Eds. (Duke University Press, Durham, North Carolina).

SBCCOM (2000). Soldier Biological Chemical Command. *Compendium of Weapons of Mass Destruction Courses* (Soldier Biological Chemical Command, Department of the Army) (http://dp.sbccom.army.mil/fr/compendium).

SHIMIZU, Y., PIERCE, D.A., PRESTON, D.L. and MABUCHI, K. (1999). "Studies of the mortality of atomic bomb survivors. Report 12, Part II. Noncancer mortality: 1950-1990," Radiat. Res. **152**, 374-389.

SIMON, J.D. (1994). *The Terrorist Trap* (Indiana Press, Bloomington, Indiana).

SLOVIC, P., LAYMAN, M. and FLYNN, J.H. (1993). "Perceived risk, trust, and nuclear waste: Lessons from Yucca Mountain," pages 64 to 88 in *Public Reactions to Nuclear Waste: Citizens' Views of Repository Siting*, Dunlap, R.E., Kraft, M.E. and Rosa, E.A., Eds. (Duke University Press, Durham, North Carolina).

STOCKWELL, H.G. and SMITH, J.M. (1998). "Involving communities in environmental health studies," pages 97 to 102 in *The Societal Aspects of Decision Making in Complex Radiological Situations* (Nuclear Energy Agency, Organisation for Economic Co-operation and Development, Paris).

STOKES, J.W. and BANDERET, L.E. (1997). "Psychological aspects of chemical defense and warfare," Military Psychol. **9**, 395-415.

TASK FORCE (1986). *Public Report of the Vice President's Task Force on Combating Terrorism, A Task Force Conclusions and Recommendations* (U.S. Government Printing Office, Washington).

TIERNEY, K.J. (1993). *Disaster Preparedness and Response: Research Findings and Guidance from the Social Science Literature*, Preliminary Paper #193 (Disaster Research Center, University of Delaware, Newark, Delaware).

UNSCEAR (2000). United Nations Scientific Committee on the Effects of Atomic Radiation. *Sources and Effects of Ionizing Radiation. UNSCEAR 2000 Report to the General Assembly, with Scientific Annexes*, No. E.00.IX.3/4 (United Nations, New York).

URSANO, R.J., FULLERTON, C.S. and MCCAUGHEY, B.G. (1994a). "Trauma and disaster," pages 3 to 27 in *Individual and Community Responses to Trauma and Disaster: The Structure of Human Chaos*, Ursano, R.J., McCaughey, B.G. and Fullerton, C.S., Eds. (Cambridge University Press, New York).

URSANO, R.J., MCCAUGHEY, B.G. and FULLERTON, C.S. (1994b). "The structure of human chaos," pages 403 to 410 in *Individual and Community Responses to Trauma and Disaster: The Structure of Human Chaos*, Ursano, R.J., McCaughey, B.G. and Fullerton, C.S., Eds. (Cambridge University Press, New York).

VYNER, H.M. (1988). "The psychological dimensions of health care for patients exposed to radiation and the other invisible environmental contaminants," Soc. Sci. Med. **27**, 1097-1103.

WANDERSMAN, A.H. and HALLMAN, W.K. (1993). "Are people acting irrationally? Understanding public concerns about environmental threats," Am. Psychol. **48**, 681-686.

YOUNG, B.H., FORD, J.D., RUZEK, J.I., FRIEDMAN, M.J. and GUSMAN, F.D. (1998). *Disaster Mental Health Services: A Guidebook for Clinicians and Administrators* (National Center for Post-Traumatic Stress Disorder, Menlo Park, California).

The NCRP

The National Council on Radiation Protection and Measurements is a nonprofit corporation chartered by Congress in 1964 to:

1. Collect, analyze, develop and disseminate in the public interest information and recommendations about (a) protection against radiation and (b) radiation measurements, quantities and units, particularly those concerned with radiation protection.
2. Provide a means by which organizations concerned with the scientific and related aspects of radiation protection and of radiation quantities, units and measurements may cooperate for effective utilization of their combined resources, and to stimulate the work of such organizations.
3. Develop basic concepts about radiation quantities, units and measurements, about the application of these concepts, and about radiation protection.
4. Cooperate with the International Commission on Radiological Protection, the International Commission on Radiation Units and Measurements, and other national and international organizations, governmental and private, concerned with radiation quantities, units and measurements and with radiation protection.

The Council is the successor to the unincorporated association of scientists known as the National Committee on Radiation Protection and Measurements and was formed to carry on the work begun by the Committee in 1929.

The participants in the Council's work are the Council members and members of scientific and administrative committees. Council members are selected solely on the basis of their scientific expertise and serve as individuals, not as representatives of any particular organization. The scientific committees, composed of experts having detailed knowledge and competence in the particular area of the committee's interest, draft proposed recommendations. These are then submitted to the full membership of the Council for careful review and approval before being published.

The following comprise the current officers and membership of the Council:

Officers

President	Charles B. Meinhold
Vice President	S. James Adelstein
Secretary and Treasurer	William M. Beckner
Assistant Secretary	Michael F. McBride

Members

S. James Adelstein
John F. Ahearne
Larry E. Anderson
Benjamin R. Archer
Mary M. Austin-Seymour
Harold L. Beck
Eleanor A. Blakely
B. Gordon Blaylock
John D. Boice, Jr.
Thomas B. Borak
André Bouville
Leslie A. Braby
David J. Brenner
Antone L. Brooks
Patricia A. Buffler
Shih-Yew Chen
Chung-Kwang Chou
James E. Cleaver
J. Donald Cossairt
Allen G. Croff
Francis A. Cucinotta
Paul M. DeLuca
Carter Denniston
Gail de Planque
John F. Dicello
Sarah S. Donaldson
William P. Dornsife
Keith F. Eckerman
Marc Edwards
Stephen A. Feig
H. Keith Florig
Kenneth R. Foster
Ethel S. Gilbert
Joel E. Gray
Andrew J. Grosovsky
Raymond A. Guilmette
William R. Hendee
David G. Hoel
F. Owen Hoffman
Geoffrey R. Howe
Kenneth R. Kase
Ann R. Kennedy
David C. Kocher
Ritsuko Komaki
Amy Kronenberg
Charles E. Land
Susan M. Langhorst
Richard W. Leggett
Howard L. Liber
James C. Lin
Jill Lipoti
John B. Little
Jay H. Lubin
C. Douglas Maynard
Claire M. Mays
Charles B. Meinhold
Fred A. Mettler, Jr.
Charles W. Miller
Jack Miller
Kenneth L. Miller
John E. Moulder
David S. Myers
Ronald C. Petersen
John W. Poston, Sr.
R. Julian Preston
Jerome S. Puskin
Genevieve S. Roessler
Marvin Rosenstein
Lawrence N. Rothenberg
Henry D. Royal
Michael T. Ryan
Jonathan M. Samet
Stephen M. Seltzer
Edward A. Sickles
David H. Sliney
Paul Slovic
Louise C. Strong
John E. Till
Lawrence W. Townsend
Lois B. Travis
Robert L. Ullrich
Richard J. Vetter
Louis K. Wagner
Daniel Wartenberg
David A. Weber
F. Ward Whicker
Chris G. Whipple
J. Frank Wilson
Susan D. Wiltshire
Marco Zaider
Pasquale Zanzonico
Marvin C. Ziskin

Honorary Members

Lauriston S. Taylor, *Honorary President,* Warren K. Sinclair, *President Emeritus*
W. Roger Ney, *Executive Director Emeritus*

Seymour Abrahamson
Edward L. Alpen
John A. Auxier
William J. Bair
Bruce B. Boecker
Victor P. Bond
Robert L. Brent
Reynold F. Brown
Melvin C. Carter
Randall S. Caswell
Frederick P. Cowan
James F. Crow
Gerald D. Dodd
Patricia W. Durbin
Thomas S. Ely
Richard F. Foster
Hymer L. Friedell
R.J. Michael Fry
Robert O. Gorson
Arthur W. Guy
Eric J. Hall
Naomi H. Harley
John W. Healy
Donald G. Jacobs
Bernd Kahn
Dade W. Moeller
A. Alan Moghissi
Robert J. Nelsen
Wesley L. Nyborg
Andrew K. Poznanski
Chester R. Richmond
William L. Russell
Eugene L. Saenger
William J. Schull
J. Newell Stannard
John B. Storer
Thomas S. Tenforde
Arthur C. Upton
George L. Voelz
Edward W. Webster

Lauriston S. Taylor Lecturers

Wesley L. Nyborg (2001) *Assuring the Safety of Medical Diagnostic Ultrasound*
S. James Adelstein (2000) *Administered Radioactivity: Unde Venimus Quoque Imus*
Naomi H. Harley (1999) *Back to Background*
Eric J. Hall (1998) *From Chimney Sweeps to Astronauts: Cancer Risks in the Workplace*
William J. Bair (1997) *Radionuclides in the Body: Meeting the Challenge!*
Seymour Abrahamson (1996) *70 Years of Radiation Genetics: Fruit Flies, Mice and Humans*
Albrecht Kellerer (1995) *Certainty and Uncertainty in Radiation Protection*
R.J. Michael Fry (1994) *Mice, Myths and Men*
Warren K. Sinclair (1993) *Science, Radiation Protection and the NCRP*
Edward W. Webster (1992) *Dose and Risk in Diagnostic Radiology: How Big? How Little?*
Victor P. Bond (1991) *When is a Dose Not a Dose?*
J. Newell Stannard (1990) *Radiation Protection and the Internal Emitter Saga*
Arthur C. Upton (1989) *Radiobiology and Radiation Protection: The Past Century and Prospects for the Future*
Bo Lindell (1988) *How Safe is Safe Enough?*
Seymour Jablon (1987) *How to be Quantitative about Radiation Risk Estimates*
Herman P. Schwan (1986) *Biological Effects of Non-ionizing Radiations: Cellular Properties and Interactions*
John H. Harley (1985) *Truth (and Beauty) in Radiation Measurement*
Harald H. Rossi (1984) *Limitation and Assessment in Radiation Protection*
Merril Eisenbud (1983) *The Human Environment—Past, Present and Future*
Eugene L. Saenger (1982) *Ethics, Trade-offs and Medical Radiation*
James F. Crow (1981) *How Well Can We Assess Genetic Risk? Not Very*
Harold O. Wyckoff (1980) *From "Quantity of Radiation" and "Dose" to "Exposure" and "Absorbed Dose"—An Historical Review*
Hymer L. Friedell (1979) *Radiation Protection—Concepts and Trade Offs*
Sir Edward Pochin (1978) *Why be Quantitative about Radiation Risk Estimates?*
Herbert M. Parker (1977) *The Squares of the Natural Numbers in Radiation Protection*

Currently, the following committees are actively engaged in formulating recommendations:

SC 1 Basic Criteria, Epidemiology, Radiobiology and Risk
SC 1-4 Extrapolation of Risks from Non-Human Experimental Systems to Man
SC 1-7 Information Needed to Make Radiation Protection Recommendations for Travel Beyond Low-Earth Orbit
SC 1-8 Risk to Thyroid from Ionizing Radiation
SC 9 Structural Shielding Design and Evaluation for Medical Use of X Rays and Gamma Rays of Energies Up to 10 MeV
SC 46 Operational Radiation Safety
SC 46-8 Radiation Protection Design Guidelines for Particle Accelerator Facilities
SC 46-10 Assessment of Occupational Doses from Internal Emitters
SC 46-13 Design of Facilities for Medical Radiation Therapy
SC 46-15 Operational Radiation Safety Program for Astronauts
SC 57-15 Uranium Risk
SC 57-17 Radionuclide Dosimetry Models for Wounds
SC 64 Environmental Issues
SC 64-19 Historical Dose
SC 64-22 Design of Effective Effluent and Environmental Monitoring Programs
SC 64-23 Cesium in the Environment
SC 66 Biological Effects and Exposure Criteria for Ultrasound
SC 72 Radiation Protection in Mammography
SC 85 Risk of Lung Cancer from Radon
SC 87 Radioactive and Mixed Waste
SC 87-1 Waste Avoidance and Volume Reduction
SC 87-2 Waste Classification Based on Risk
SC 87-3 Performance Assessment
SC 87-4 Management of Waste Metals Containing Radioactivity
SC 89 Nonionizing Electromagnetic Fields
SC 89-3 Biological Effects of Extremely Low-Frequency Electric and Magnetic Fields
SC 89-4 Biological Effects and Exposure Recommendations for Modulated Radiofrequency Fields
SC 91 Radiation Protection in Medicine
SC 91-1 Precautions in the Management of Patients Who Have Received Therapeutic Amounts of Radionuclides
SC 91-2 Radiation Protection in Dentistry
SC 92 Public Policy and Risk Communication
SC 93 Radiation Measurement and Dosimetry

In recognition of its responsibility to facilitate and stimulate cooperation among organizations concerned with the scientific and related aspects of radiation protection and measurement, the Council has created a category of NCRP Collaborating Organizations. Organizations or groups of organizations that are national or international in scope and are concerned with scientific problems involving radiation quantities, units, measurements and effects, or radiation protection may be admitted to collaborating status by the Council. Collaborating Organizations provide a means by which the NCRP can gain input into its activities from a wider segment of society. At the same time, the relationships with the Collaborating Organizations facilitate wider dissemination of information about the Council's activities, interests and concerns. Collaborating Organizations have the opportunity to comment on draft reports (at the time that these are submitted to the members of the Council). This is intended to capitalize on the fact that Collaborating Organizations are in an excellent position to both contribute to the identification of what needs to be treated in NCRP reports and to identify problems that might result from proposed recommendations. The present Collaborating Organizations with which the NCRP maintains liaison are as follows:

Agency for Toxic Substances and Disease Registry
American Academy of Dermatology
American Academy of Environmental Engineers
American Academy of Health Physics
American Association of Physicists in Medicine
American College of Medical Physics
American College of Nuclear Physicians
American College of Occupational and Environmental Medicine
American College of Radiology
American Dental Association
American Industrial Hygiene Association
American Institute of Ultrasound in Medicine
American Insurance Services Group
American Medical Association
American Nuclear Society
American Pharmaceutical Association
American Podiatric Medical Association
American Public Health Association
American Radium Society
American Roentgen Ray Society
American Society for Therapeutic Radiology and Oncology
American Society of Health-System Pharmacists
American Society of Radiologic Technologists
Association of University Radiologists
Bioelectromagnetics Society
Campus Radiation Safety Officers
College of American Pathologists
Conference of Radiation Control Program Directors, Inc.
Council on Radionuclides and Radiopharmaceuticals
Defense Threat Reduction Agency
Electric Power Research Institute
Electromagnetic Energy Association

Federal Communications Commission
Federal Emergency Management Agency
Genetics Society of America
Health Physics Society
Institute of Electrical and Electronics Engineers, Inc.
Institute of Nuclear Power Operations
International Brotherhood of Electrical Workers
National Aeronautics and Space Administration
National Association of Environmental Professionals
National Electrical Manufacturers Association
National Institute for Occupational Safety and Health
National Institute of Standards and Technology
Nuclear Energy Institute
Office of Science and Technology Policy
Paper, Allied-Industrial, Chemical and Energy Workers International Union
Radiation Research Society
Radiological Society of North America
Society for Risk Analysis
Society of Nuclear Medicine
U.S. Air Force
U.S. Army
U.S. Coast Guard
U.S. Department of Energy
U.S. Department of Housing and Urban Development
U.S. Department of Labor
U.S. Department of Transportation
U.S. Environmental Protection Agency
U.S. Navy
U.S. Nuclear Regulatory Commission
U.S. Public Health Service
Utility Workers Union of America

The NCRP has found its relationships with these organizations to be extremely valuable to continued progress in its program.

Another aspect of the cooperative efforts of the NCRP relates to the Special Liaison relationships established with various governmental organizations that have an interest in radiation protection and measurements. This liaison relationship provides: (1) an opportunity for participating organizations to designate an individual to provide liaison between the organization and the NCRP; (2) that the individual designated will receive copies of draft NCRP reports (at the time that these are submitted to the members of the Council) with an invitation to comment, but not vote; and (3) that new NCRP efforts might be discussed with liaison individuals as appropriate, so that they might have an opportunity to make suggestions on new studies and related matters. The following organizations participate in the Special Liaison Program:

Atomic Energy Control Board
Australian Radiation Laboratory
Bundesamt für Strahlenschutz (Germany)
Central Laboratory for Radiological Protection (Poland)
China Institute for Radiation Protection
Commisariat à l'Energie Atomique
Commonwealth Scientific Instrumentation Research Organization (Australia)
European Commission
Health Council of the Netherlands
International Commission on Non-Ionizing Radiation Protection
Japan Radiation Council
Korea Institute of Nuclear Safety
National Radiological Protection Board (United Kingdom)
Russian Scientific Commission on Radiation Protection
South African Forum for Radiation Protection
World Association of Nuclear Operations

The NCRP values highly the participation of these organizations in the Special Liaison Program.

The Council also benefits significantly from the relationships established pursuant to the Corporate Sponsor's Program. The program facilitates the interchange of information and ideas and corporate sponsors provide valuable fiscal support for the Council's program. This developing program currently includes the following Corporate Sponsors:

3M Corporate Health Physics
Commonwealth Edison
Consolidated Edison
Duke Energy Corporation
Florida Power Corporation
ICN Biomedicals, Inc.
Landauer, Inc.
New York Power Authority
Nuclear Energy Institute
Nycomed Amersham Corporation
Southern California Edison

The Council's activities are made possible by the voluntary contribution of time and effort by its members and participants and the generous support of the following organizations:

3M Health Physics Services
Agfa Corporation

Alfred P. Sloan Foundation
Alliance of American Insurers
American Academy of Dermatology
American Academy of Health Physics
American Academy of Oral and Maxillofacial Radiology
American Association of Physicists in Medicine
American Cancer Society
American College of Medical Physics
American College of Nuclear Physicians
American College of Occupational and Environmental Medicine
American College of Radiology
American College of Radiology Foundation
American Dental Association
American Healthcare Radiology Administrators
American Industrial Hygiene Association
American Insurance Services Group
American Medical Association
American Nuclear Society
American Osteopathic College of Radiology
American Podiatric Medical Association
American Public Health Association
American Radium Society
American Roentgen Ray Society
American Society of Radiologic Technologists
American Society for Therapeutic Radiology and Oncology
American Veterinary Medical Association
American Veterinary Radiology Society
Association of University Radiologists
Battelle Memorial Institute
Canberra Industries, Inc.
Chem Nuclear Systems
Center for Devices and Radiological Health
College of American Pathologists
Committee on Interagency Radiation Research
 and Policy Coordination
Commonwealth of Pennsylvania
Consumers Power Company
Council on Radionuclides and Radiopharmaceuticals
Defense Nuclear Agency
Eastman Kodak Company
Edison Electric Institute
Edward Mallinckrodt, Jr. Foundation
EG&G Idaho, Inc.
Electric Power Research Institute
Federal Emergency Management Agency
Florida Institute of Phosphate Research
Fuji Medical Systems, U.S.A., Inc.

Genetics Society of America
Health Effects Research Foundation (Japan)
Health Physics Society
Institute of Nuclear Power Operations
James Picker Foundation
Martin Marietta Corporation
Motorola Foundation
National Aeronautics and Space Administration
National Association of Photographic Manufacturers
National Cancer Institute
National Electrical Manufacturers Association
National Institute of Standards and Technology
Picker International
Public Service Electric and Gas Company
Radiation Research Society
Radiological Society of North America
Richard Lounsbery Foundation
Sandia National Laboratory
Siemens Medical Systems, Inc.
Society of Nuclear Medicine
Society of Pediatric Radiology
U.S. Department of Energy
U.S. Department of Labor
U.S. Environmental Protection Agency
U.S. Navy
U.S. Nuclear Regulatory Commission
Victoreen, Inc.
Westinghouse Electric Corporation

Initial funds for publication of NCRP reports were provided by a grant from the James Picker Foundation.

The NCRP seeks to promulgate information and recommendations based on leading scientific judgment on matters of radiation protection and measurement and to foster cooperation among organizations concerned with these matters. These efforts are intended to serve the public interest and the Council welcomes comments and suggestions on its reports or activities from those interested in its work.

NCRP Publications

Information on NCRP publications may be obtained from the NCRP website (http://www.ncrp.com) or by telephone (800-229-2652) and fax (301-907-8768). The address is:

NCRP Publications
7910 Woodmont Avenue
Suite 800
Bethesda, MD 20814-3095

Abstracts of NCRP reports published since 1980, abstracts of all NCRP commentaries, and the text of all NCRP statements are available at the NCRP website. Currently available publications are listed below.

NCRP Reports

No.	Title
8	*Control and Removal of Radioactive Contamination in Laboratories* (1951)
22	*Maximum Permissible Body Burdens and Maximum Permissible Concentrations of Radionuclides in Air and in Water for Occupational Exposure* (1959) [includes Addendum 1 issued in August 1963]
25	*Measurement of Absorbed Dose of Neutrons, and of Mixtures of Neutrons and Gamma Rays* (1961)
27	*Stopping Powers for Use with Cavity Chambers* (1961)
30	*Safe Handling of Radioactive Materials* (1964)
32	*Radiation Protection in Educational Institutions* (1966)
35	*Dental X-Ray Protection* (1970)
36	*Radiation Protection in Veterinary Medicine* (1970)
37	*Precautions in the Management of Patients Who Have Received Therapeutic Amounts of Radionuclides* (1970)
38	*Protection Against Neutron Radiation* (1971)
40	*Protection Against Radiation from Brachytherapy Sources* (1972)
41	*Specification of Gamma-Ray Brachytherapy Sources* (1974)
42	*Radiological Factors Affecting Decision-Making in a Nuclear Attack* (1974)

44 *Krypton-85 in the Atmosphere—Accumulation, Biological Significance, and Control Technology* (1975)
46 *Alpha-Emitting Particles in Lungs* (1975)
47 *Tritium Measurement Techniques* (1976)
49 *Structural Shielding Design and Evaluation for Medical Use of X Rays and Gamma Rays of Energies Up to 10 MeV* (1976)
50 *Environmental Radiation Measurements* (1976)
51 *Radiation Protection Design Guidelines for 0.1-10 MeV Particle Accelerator Facilities* (1977)
52 *Cesium-137 from the Environment to Man: Metabolism and Dose* (1977)
54 *Medical Radiation Exposure of Pregnant and Potentially Pregnant Women* (1977)
55 *Protection of the Thyroid Gland in the Event of Releases of Radioiodine* (1977)
57 *Instrumentation and Monitoring Methods for Radiation Protection* (1978)
58 *A Handbook of Radioactivity Measurements Procedures*, 2nd ed. (1985)
59 *Operational Radiation Safety Program* (1978)
60 *Physical, Chemical, and Biological Properties of Radiocerium Relevant to Radiation Protection Guidelines* (1978)
61 *Radiation Safety Training Criteria for Industrial Radiography* (1978)
62 *Tritium in the Environment* (1979)
63 *Tritium and Other Radionuclide Labeled Organic Compounds Incorporated in Genetic Material* (1979)
64 *Influence of Dose and Its Distribution in Time on Dose-Response Relationships for Low-LET Radiations* (1980)
65 *Management of Persons Accidentally Contaminated with Radionuclides* (1980)
67 *Radiofrequency Electromagnetic Fields—Properties, Quantities and Units, Biophysical Interaction, and Measurements* (1981)
68 *Radiation Protection in Pediatric Radiology* (1981)
69 *Dosimetry of X-Ray and Gamma-Ray Beams for Radiation Therapy in the Energy Range 10 keV to 50 MeV* (1981)
70 *Nuclear Medicine—Factors Influencing the Choice and Use of Radionuclides in Diagnosis and Therapy* (1982)
72 *Radiation Protection and Measurement for Low-Voltage Neutron Generators* (1983)
73 *Protection in Nuclear Medicine and Ultrasound Diagnostic Procedures in Children* (1983)
74 *Biological Effects of Ultrasound: Mechanisms and Clinical Implications* (1983)
75 *Iodine-129: Evaluation of Releases from Nuclear Power Generation* (1983)

77 *Exposures from the Uranium Series with Emphasis on Radon and Its Daughters* (1984)
78 *Evaluation of Occupational and Environmental Exposures to Radon and Radon Daughters in the United States* (1984)
79 *Neutron Contamination from Medical Electron Accelerators* (1984)
80 *Induction of Thyroid Cancer by Ionizing Radiation* (1985)
81 *Carbon-14 in the Environment* (1985)
82 *SI Units in Radiation Protection and Measurements* (1985)
83 *The Experimental Basis for Absorbed-Dose Calculations in Medical Uses of Radionuclides* (1985)
84 *General Concepts for the Dosimetry of Internally Deposited Radionuclides* (1985)
85 *Mammography—A User's Guide* (1986)
86 *Biological Effects and Exposure Criteria for Radiofrequency Electromagnetic Fields* (1986)
87 *Use of Bioassay Procedures for Assessment of Internal Radionuclide Deposition* (1987)
88 *Radiation Alarms and Access Control Systems* (1986)
89 *Genetic Effects from Internally Deposited Radionuclides* (1987)
90 *Neptunium: Radiation Protection Guidelines* (1988)
92 *Public Radiation Exposure from Nuclear Power Generation in the United States* (1987)
93 *Ionizing Radiation Exposure of the Population of the United States* (1987)
94 *Exposure of the Population in the United States and Canada from Natural Background Radiation* (1987)
95 *Radiation Exposure of the U.S. Population from Consumer Products and Miscellaneous Sources* (1987)
96 *Comparative Carcinogenicity of Ionizing Radiation and Chemicals* (1989)
97 *Measurement of Radon and Radon Daughters in Air* (1988)
98 *Guidance on Radiation Received in Space Activities* (1989)
99 *Quality Assurance for Diagnostic Imaging* (1988)
100 *Exposure of the U.S. Population from Diagnostic Medical Radiation* (1989)
101 *Exposure of the U.S. Population from Occupational Radiation* (1989)
102 *Medical X-Ray, Electron Beam and Gamma-Ray Protection for Energies Up to 50 MeV (Equipment Design, Performance and Use)* (1989)
103 *Control of Radon in Houses* (1989)
104 *The Relative Biological Effectiveness of Radiations of Different Quality* (1990)
105 *Radiation Protection for Medical and Allied Health Personnel* (1989)
106 *Limit for Exposure to "Hot Particles" on the Skin* (1989)

107 *Implementation of the Principle of As Low As Reasonably Achievable (ALARA) for Medical and Dental Personnel* (1990)
108 *Conceptual Basis for Calculations of Absorbed-Dose Distributions* (1991)
109 *Effects of Ionizing Radiation on Aquatic Organisms* (1991)
110 *Some Aspects of Strontium Radiobiology* (1991)
111 *Developing Radiation Emergency Plans for Academic, Medical or Industrial Facilities* (1991)
112 *Calibration of Survey Instruments Used in Radiation Protection for the Assessment of Ionizing Radiation Fields and Radioactive Surface Contamination* (1991)
113 *Exposure Criteria for Medical Diagnostic Ultrasound: I. Criteria Based on Thermal Mechanisms* (1992)
114 *Maintaining Radiation Protection Records* (1992)
115 *Risk Estimates for Radiation Protection* (1993)
116 *Limitation of Exposure to Ionizing Radiation* (1993)
117 *Research Needs for Radiation Protection* (1993)
118 *Radiation Protection in the Mineral Extraction Industry* (1993)
119 *A Practical Guide to the Determination of Human Exposure to Radiofrequency Fields* (1993)
120 *Dose Control at Nuclear Power Plants* (1994)
121 *Principles and Application of Collective Dose in Radiation Protection* (1995)
122 *Use of Personal Monitors to Estimate Effective Dose Equivalent and Effective Dose to Workers for External Exposure to Low-LET Radiation* (1995)
123 *Screening Models for Releases of Radionuclides to Atmosphere, Surface Water, and Ground* (1996)
124 *Sources and Magnitude of Occupational and Public Exposures from Nuclear Medicine Procedures* (1996)
125 *Deposition, Retention and Dosimetry of Inhaled Radioactive Substances* (1997)
126 *Uncertainties in Fatal Cancer Risk Estimates Used in Radiation Protection* (1997)
127 *Operational Radiation Safety Program* (1998)
128 *Radionuclide Exposure of the Embryo / Fetus* (1998)
129 *Recommended Screening Limits for Contaminated Surface Soil and Review of Factors Relevant to Site-Specific Studies* (1999)
130 *Biological Effects and Exposure Limits for "Hot Particles"* (1999)
131 *Scientific Basis for Evaluating the Risks to Populations from Space Applications of Plutonium* (2001)
132 *Radiation Protection Guidance for Activities in Low-Earth Orbit* (2000)
133 *Radiation Protection for Procedures Performed Outside the Radiology Department* (2000)
134 *Operational Radiation Safety Training* (2000)
135 *Liver Cancer Risk from Internally-Deposited Radionuclides* (2001)

136 *Evaluation of the Linear-Nonthreshold Dose-Response Model for Ionizing Radiation* (2001)
137 *Fluence-Based and Microdosimetric Event-Based Methods for Radiation Protection in Space* (2001)
138 *Management of Terrorist Events Involving Radioactive Material* (2001)

Binders for NCRP reports are available. Two sizes make it possible to collect into small binders the "old series" of reports (NCRP Reports Nos. 8–30) and into large binders the more recent publications (NCRP Reports Nos. 32–138). Each binder will accommodate from five to seven reports. The binders carry the identification "NCRP Reports" and come with label holders which permit the user to attach labels showing the reports contained in each binder.

The following bound sets of NCRP reports are also available:

Volume I. NCRP Reports Nos. 8, 22
Volume II. NCRP Reports Nos. 23, 25, 27, 30
Volume III. NCRP Reports Nos. 32, 35, 36, 37
Volume IV. NCRP Reports Nos. 38, 40, 41
Volume V. NCRP Reports Nos. 42, 44, 46
Volume VI. NCRP Reports Nos. 47, 49, 50, 51
Volume VII. NCRP Reports Nos. 52, 53, 54, 55, 57
Volume VIII. NCRP Report No. 58
Volume IX. NCRP Reports Nos. 59, 60, 61, 62, 63
Volume X. NCRP Reports Nos. 64, 65, 66, 67
Volume XI. NCRP Reports Nos. 68, 69, 70, 71, 72
Volume XII. NCRP Reports Nos. 73, 74, 75, 76
Volume XIII. NCRP Reports Nos. 77, 78, 79, 80
Volume XIV. NCRP Reports Nos. 81, 82, 83, 84, 85
Volume XV. NCRP Reports Nos. 86, 87, 88, 89
Volume XVI. NCRP Reports Nos. 90, 91, 92, 93
Volume XVII. NCRP Reports Nos. 94, 95, 96, 97
Volume XVIII. NCRP Reports Nos. 98, 99, 100
Volume XIX. NCRP Reports Nos. 101, 102, 103, 104
Volume XX. NCRP Reports Nos. 105, 106, 107, 108
Volume XXI. NCRP Reports Nos. 109, 110, 111
Volume XXII. NCRP Reports Nos. 112, 113, 114
Volume XXIII. NCRP Reports Nos. 115, 116, 117, 118
Volume XXIV. NCRP Reports Nos. 119, 120, 121, 122
Volume XXV. NCRP Report No. 123I and 123II
Volume XXVI. NCRP Reports Nos. 124, 125, 126, 127
Volume XXVII. NCRP Reports Nos. 128, 129, 130

(Titles of the individual reports contained in each volume are given previously.)

NCRP Commentaries

No.	Title
1	*Krypton-85 in the Atmosphere—With Specific Reference to the Public Health Significance of the Proposed Controlled Release at Three Mile Island* (1980)
4	*Guidelines for the Release of Waste Water from Nuclear Facilities with Special Reference to the Public Health Significance of the Proposed Release of Treated Waste Waters at Three Mile Island* (1987)
5	*Review of the Publication, Living Without Landfills* (1989)
6	*Radon Exposure of the U.S. Population—Status of the Problem* (1991)
7	*Misadministration of Radioactive Material in Medicine—Scientific Background* (1991)
8	*Uncertainty in NCRP Screening Models Relating to Atmospheric Transport, Deposition and Uptake by Humans* (1993)
9	*Considerations Regarding the Unintended Radiation Exposure of the Embryo, Fetus or Nursing Child* (1994)
10	*Advising the Public about Radiation Emergencies: A Document for Public Comment* (1994)
11	*Dose Limits for Individuals Who Receive Exposure from Radionuclide Therapy Patients* (1995)
12	*Radiation Exposure and High-Altitude Flight* (1995)
13	*An Introduction to Efficacy in Diagnostic Radiology and Nuclear Medicine (Justification of Medical Radiation Exposure)* (1995)
14	*A Guide for Uncertainty Analysis in Dose and Risk Assessments Related to Environmental Contamination* (1996)
15	*Evaluating the Reliability of Biokinetic and Dosimetric Models and Parameters Used to Assess Individual Doses for Risk Assessment Purposes* (1998)

Proceedings of the Annual Meeting

No.	Title
1	*Perceptions of Risk*, Proceedings of the Fifteenth Annual Meeting held on March 14-15, 1979 (including Taylor Lecture No. 3) (1980)
3	*Critical Issues in Setting Radiation Dose Limits*, Proceedings of the Seventeenth Annual Meeting held on April 8-9, 1981 (including Taylor Lecture No. 5) (1982)

4 *Radiation Protection and New Medical Diagnostic Approaches*, Proceedings of the Eighteenth Annual Meeting held on April 6-7, 1982 (including Taylor Lecture No. 6) (1983)

5 *Environmental Radioactivity*, Proceedings of the Nineteenth Annual Meeting held on April 6-7, 1983 (including Taylor Lecture No. 7) (1983)

6 *Some Issues Important in Developing Basic Radiation Protection Recommendations*, Proceedings of the Twentieth Annual Meeting held on April 4-5, 1984 (including Taylor Lecture No. 8) (1985)

7 *Radioactive Waste*, Proceedings of the Twenty-first Annual Meeting held on April 3-4, 1985 (including Taylor Lecture No. 9)(1986)

8 *Nonionizing Electromagnetic Radiations and Ultrasound*, Proceedings of the Twenty-second Annual Meeting held on April 2-3, 1986 (including Taylor Lecture No. 10) (1988)

9 *New Dosimetry at Hiroshima and Nagasaki and Its Implications for Risk Estimates*, Proceedings of the Twenty-third Annual Meeting held on April 8-9, 1987 (including Taylor Lecture No. 11) (1988)

10 *Radon*, Proceedings of the Twenty-fourth Annual Meeting held on March 30-31, 1988 (including Taylor Lecture No. 12) (1989)

11 *Radiation Protection Today—The NCRP at Sixty Years*, Proceedings of the Twenty-fifth Annual Meeting held on April 5-6, 1989 (including Taylor Lecture No. 13) (1990)

12 *Health and Ecological Implications of Radioactively Contaminated Environments*, Proceedings of the Twenty-sixth Annual Meeting held on April 4-5, 1990 (including Taylor Lecture No. 14) (1991)

13 *Genes, Cancer and Radiation Protection*, Proceedings of the Twenty-seventh Annual Meeting held on April 3-4, 1991 (including Taylor Lecture No. 15) (1992)

14 *Radiation Protection in Medicine*, Proceedings of the Twenty-eighth Annual Meeting held on April 1-2, 1992 (including Taylor Lecture No. 16) (1993)

15 *Radiation Science and Societal Decision Making*, Proceedings of the Twenty-ninth Annual Meeting held on April 7-8, 1993 (including Taylor Lecture No. 17) (1994)

16 *Extremely-Low-Frequency Electromagnetic Fields: Issues in Biological Effects and Public Health*, Proceedings of the Thirtieth Annual Meeting held on April 6-7, 1994 (not published).

17 *Environmental Dose Reconstruction and Risk Implications*, Proceedings of the Thirty-first Annual Meeting held on April 12-13, 1995 (including Taylor Lecture No. 19) (1996)

18 *Implications of New Data on Radiation Cancer Risk*, Proceedings of the Thirty-second Annual Meeting held on April 3-4, 1996 (including Taylor Lecture No. 20) (1997)
19 *The Effects of Pre- and Postconception Exposure to Radiation*, Proceedings of the Thirty-third Annual Meeting held on April 2-3, 1997, Teratology **59**, 181-317 (1999)
20 *Cosmic Radiation Exposure of Airline Crews, Passengers and Astronauts*, Proceedings of the Thirty-fourth Annual Meeting held on April 1-2, 1998, Health Phys. **79**, 466-613 (2000)
21 *Radiation Protection in Medicine: Contemporary Issues*, Proceedings of the Thirty-fifth Annual Meeting held on April 7-8, 1999 (including Taylor Lecture No. 23) (1999)
22 *Ionizing Radiation Science and Protection in the 21st Century*, Proceedings of the Thirty-sixth Annual Meeting held on April 5-6, 2000, Health Phys. **80**, 317-402 (2001)

Lauriston S. Taylor Lectures

No.	Title
1	*The Squares of the Natural Numbers in Radiation Protection* by Herbert M. Parker (1977)
2	*Why be Quantitative about Radiation Risk Estimates?* by Sir Edward Pochin (1978)
3	*Radiation Protection—Concepts and Trade Offs* by Hymer L. Friedell (1979) [available also in *Perceptions of Risk*, see above]
4	*From "Quantity of Radiation" and "Dose" to "Exposure" and "Absorbed Dose"—An Historical Review* by Harold O. Wyckoff (1980)
5	*How Well Can We Assess Genetic Risk? Not Very* by James F. Crow (1981) [available also in *Critical Issues in Setting Radiation Dose Limits*, see above]
6	*Ethics, Trade-offs and Medical Radiation* by Eugene L. Saenger (1982) [available also in *Radiation Protection and New Medical Diagnostic Approaches*, see above]
7	*The Human Environment—Past, Present and Future* by Merril Eisenbud (1983) [available also in *Environmental Radioactivity*, see above]
8	*Limitation and Assessment in Radiation Protection* by Harald H. Rossi (1984) [available also in *Some Issues Important in Developing Basic Radiation Protection Recommendations*, see above]
9	*Truth (and Beauty) in Radiation Measurement* by John H. Harley (1985) [available also in *Radioactive Waste*, see above]
10	*Biological Effects of Non-ionizing Radiations: Cellular Properties and Interactions* by Herman P. Schwan (1987) [available also in

Nonionizing Electromagnetic Radiations and Ultrasound, see above]

11 *How to be Quantitative about Radiation Risk Estimates* by Seymour Jablon (1988) [available also in *New Dosimetry at Hiroshima and Nagasaki and its Implications for Risk Estimates*, see above]

12 *How Safe is Safe Enough?* by Bo Lindell (1988) [available also in *Radon*, see above]

13 *Radiobiology and Radiation Protection: The Past Century and Prospects for the Future* by Arthur C. Upton (1989) [available also in *Radiation Protection Today*, see above]

14 *Radiation Protection and the Internal Emitter Saga* by J. Newell Stannard (1990) [available also in *Health and Ecological Implications of Radioactively Contaminated Environments*, see above]

15 *When is a Dose Not a Dose?* by Victor P. Bond (1992) [available also in *Genes, Cancer and Radiation Protection*, see above]

16 *Dose and Risk in Diagnostic Radiology: How Big? How Little?* by Edward W. Webster (1992)[available also in *Radiation Protection in Medicine*, see above]

17 *Science, Radiation Protection and the NCRP* by Warren K. Sinclair (1993)[available also in *Radiation Science and Societal Decision Making*, see above]

18 *Mice, Myths and Men* by R.J. Michael Fry (1995)

19 *Certainty and Uncertainty in Radiation Research* by Albrecht M. Kellerer (1995). Health Phys. **69**, 446–453.

20 *70 Years of Radiation Genetics: Fruit Flies, Mice and Humans* by Seymour Abrahamson (1996). Health Phys. **71**, 624–633.

21 *Radionuclides in the Body: Meeting the Challenge* by William J. Bair (1997). Health Phys. **73**, 423–432.

22 *From Chimney Sweeps to Astronauts: Cancer Risks in the Work Place* by Eric J. Hall (1998). Health Phys. **75**, 357–366.

23 *Back to Background: Natural Radiation and Radioactivity Exposed* by Naomi H. Harley (2000). Health Phys. **79**, 121–128.

24 *Administered Radioactivity: Unde Venimus Quoque Imus* by S. James Adelstein (2001). Health Phys. **80**, 317-324.

Symposium Proceedings

No.	Title
1	*The Control of Exposure of the Public to Ionizing Radiation in the Event of Accident or Attack*, Proceedings of a Symposium held April 27-29, 1981 (1982)

2	*Radioactive and Mixed Waste—Risk as a Basis for Waste Classification,* Proceedings of a Symposium held November 9, 1994 (1995)
3	*Acceptability of Risk from Radiation—Application to Human Space Flight,* Proceedings of a Symposium held May 29, 1996 (1997)

NCRP Statements

No.	Title
1	"Blood Counts, Statement of the National Committee on Radiation Protection," Radiology **63**, 428 (1954)
2	"Statements on Maximum Permissible Dose from Television Receivers and Maximum Permissible Dose to the Skin of the Whole Body," Am. J. Roentgenol., Radium Ther. and Nucl. Med. **84**, 152 (1960) and Radiology **75**, 122 (1960)
3	*X-Ray Protection Standards for Home Television Receivers, Interim Statement of the National Council on Radiation Protection and Measurements* (1968)
4	*Specification of Units of Natural Uranium and Natural Thorium, Statement of the National Council on Radiation Protection and Measurements* (1973)
5	*NCRP Statement on Dose Limit for Neutrons* (1980)
6	*Control of Air Emissions of Radionuclides* (1984)
7	*The Probability That a Particular Malignancy May Have Been Caused by a Specified Irradiation* (1992)
8	*The Application of ALARA for Occupational Exposures* (1999)
9	*Extension of the Skin Dose Limit for Hot Particles to Other External Sources of Skin Irradiation* (2001)

Other Documents

The following documents of the NCRP were published outside of the NCRP report, commentary and statement series:

Somatic Radiation Dose for the General Population, Report of the Ad Hoc Committee of the National Council on Radiation Protection and Measurements, 6 May 1959, Science, February 19, 1960, Vol. 131, No. 3399, pages 482-486

Dose Effect Modifying Factors in Radiation Protection, Report of Subcommittee M-4 (Relative Biological Effectiveness) of the National Council on Radiation Protection and Measurements, Report BNL 50073 (T-471) (1967) Brookhaven National Laboratory (National Technical Information Service, Springfield, Virginia)

Index